MW01283228

Astrophysics of Exoplanetary Atmospheres

Volume 450

Astrophysics and Space Science Library

More information about this series at http://www.springer.com/series/5664

Valerio Bozza · Luigi Mancini
Alessandro Sozzetti
Editors

Astrophysics of Exoplanetary Atmospheres

2nd Advanced School on Exoplanetary Science

Editors
Valerio Bozza
Department of Physics
University of Salerno
Fisciano
Italy

Alessandro Sozzetti
INAF–Osservatorio Astrofisico di Torino
Pino Torinese, Turin
Italy

Luigi Mancini
Department of Physics
University of Rome Tor Vergata
Roma
Italy

ISSN 0067-0057 ISSN 2214-7985 (electronic)
Astrophysics and Space Science Library
ISBN 978-3-319-89700-4 ISBN 978-3-319-89701-1 (eBook)
https://doi.org/10.1007/978-3-319-89701-1

Library of Congress Control Number: 2018938779

Cover illustration: Faint Signatures of Water in Exoplanet's Atmosphere. Credit NASA Goddard Space Flight Center; NASA/STScI-2013-54, ESA, A. Mandell (Goddard Space Flight Center), and D. Deming (University of Maryland, College Park)

Printed on acid-free paper

This Springer imprint is published by the registered company Springer International Publishing AG part of Springer Nature
The registered company address is: Gewerbestrasse 11, 6330 Cham, Switzerland

To our old ladies

Preface

The mass of the Earth's atmosphere constitutes only 1 millionth of the total mass of the planet, but without it, we would not be here to write the Preface of this book. Indeed, the development of complex life (as we know it) and its long-term sustainability on the surface of a rocky planet with abundant reservoirs of liquid water are inextricably tied to the presence of a planetary atmosphere. In the search for an answer to the age-old question of mankind: "Are we alone?" it is thus clear that the study and understanding of the atmospheres of extrasolar planets are an unavoidable step toward the goal. However, this is no easy task: The field of exoplanetary atmospheres is highly interdisciplinary, and it requires the application of knowledge from astronomy and astrophysics, atmospheric and climate science, chemistry, geology and geophysics, planetary science, and even biology and quantum physics.

The announcement of the first detection of an atomic species (sodium) in the atmosphere of a planet orbiting a star other than the Sun is now 15 years old. It heralded the beginning of an era of extraordinary measurements in exoplanetary science, which exploited for the most part the application of cunning techniques for the investigation of the class of transiting exoplanets. Nature's infinite creativity in forming worlds in a continuum of physical (masses, radii) and architectural properties (orbital shapes and separations, orbital alignment and multiplicity), spectacularly exceeding the variety represented by our Solar System, suggests that we should expect no less from the possible realizations of exoplanetary atmospheres. For a decade, the difficulty in carrying out robust spectroscopic measurements of atomic and molecular species has often limited the depth in the interpretation of the properties of the atmospheres of exoplanets by comparison with theoretical spectra. But the most recent, systematic studies with the best available space-borne instrumentation are beginning to unveil precisely the same continuum in the variety of atmospheric properties, particularly for the class of transiting strongly irradiated giant planets. As the era of the James Webb Space Telescope and of the ground-based Extremely Large Telescopes is about to unfold, the prospects appear bright for tackling vast observational studies of the atmospheres of Neptune- and Super-Earth-type exoplanets, even at habitable zone distances from their parent stars. Observations alone, however, are not enough—theoretical advances to aid in the interpretation of the observations and to guide future choices in observing strategies will be a key. Indeed, the tools used to model exoplanetary atmospheres are often the very same tools, or an adaption of the tools, employed to model the atmospheres of the planets in the Solar System. This, on the one hand, will ease future studies of comparative planetology. On the other hand, the complex interplay between the atmospheric processes governed by the physics of radiation, fluid dynamics, atmospheric chemistry, and atmospheric escape is bound to provide interpretative challenges, when it is seen at work in objects with entirely different properties (composition, irradiation conditions) from those of the Solar System planets. Furthermore, improved modeling, interpretation, and understanding of the extremely exotic atmospheres of worlds orbiting other stars (and their environment) critically depend on laboratory data to deliver necessary physical and chemical inputs and constraints, but there are many critical areas in which laboratory work still needs to fill important gaps.

With the ensuing tide of forthcoming observations bound to revolutionize our ability to characterize the atmospheric structure, composition, and circulation of extrasolar worlds, it appeared entirely fitting and timely to focus the 2nd Advanced School on Exoplanetary Science on the astrophysics of extrasolar planets' atmospheres. This book captures the details of the lectures given by four world-class experts in this vast field. While it has no ambition of being exhaustive, the material in this book has been organized so as to provide, for the first time, a single reference source describing, in a long exposure still picture, the state of the art of some of the most relevant areas of work in exoplanetary atmospheres, both from an observational, theoretical, and laboratory perspective, and with an eye at the direct comparison with our knowledge of the atmospheric properties of Solar System planets. First, atmospheric measurements for exoplanets are presented using the class of transiting systems as benchmark. The overall background and methodology of the measurements aimed at effectively separating the planetary signals from that of its host star are described, together with analysis and fitting methods, for the three fundamental techniques of transmission (during primary transit events) and emission (during secondary eclipse events) spectroscopy, and photometry along the planet's orbit (phase curves). The determination of the observable properties of exoplanetary atmospheres is then connected to some of the major scientific questions, which range from broad issues about planet formation and migration, to detailed atmospheric physics questions about how a planet's atmosphere responds under extreme conditions. Next, the most relevant elements of the calculation of present-day exoplanet model atmospheres are reviewed, focusing on the morphology of the emergent spectra based on our understanding of atmospheric temperature–pressure profiles. The atmospheric models of both strongly irradiated and isolated gas giant planets are pivotal in the discussion, but connections are made to the well-studied atmospheres of brown dwarfs as well as sub-Neptunes and terrestrial planets. Several important modeling aspects are discussed, which include the role of stellar irradiation, chemical composition, surface gravity, atmospheric abundances, interior fluxes, and cloud opacity, with illustrative examples of model atmosphere retrievals on a thermal emission spectrum and outlined connections to the predictions of planet formation models. Then, a survey is presented of the (quantum mechanical) rules governing the spectroscopy of atoms and molecules likely to be present in the atmospheres of extrasolar planets, using illustrative examples based on the comparison with observed spectra. The proper understanding of the spectroscopic signatures of atomic and molecular species in exoplanetary atmospheres requires a wealth of laboratory data to characterize those species, which in turn serve as critical inputs to atmospheric physical and chemical models. The basics of atomic and molecular spectroscopy are presented with a particular emphasis on the importance of treating the temperature dependence of the spectrum and the huge growth in the number of lines which play a role at higher temperatures, such as those deduced for transiting hot Jupiters, the best studied sample to date. Sources of data (spectroscopic databases) for use in studies of exoplanets are also discussed in detail, and illustrative examples are given. Finally, an overview of the atmospheric physics and atmospheres of Solar System bodies is presented. Solar System studies have already shown us a need to understand atmospheric processes in temperature regimes (both far hotter and colder) entirely different from that of the Earth, and with similarly vast differences in compositions. The properties of the atmospheres in the Solar System thus constitute a set of important comparison metrics to be used to further our understanding of the seemingly exotic regimes of atmospheres found in extrasolar planets. The fundamental processes governing the origin, structure, and evolution (both physical and chemical) of the atmospheres of Solar System bodies are discussed, along with basic concepts of atmospheric dynamics. A survey of the atmospheres of the Solar System is then presented, focusing on the bodies that retain most of their atmosphere in the collisional regime (terrestrial planets, giant planets, and icy bodies such as the moons of giant planets and trans-Neptunian objects).

This book provides an essential introduction to the observational, theoretical, and laboratory sides of the field of exoplanetary atmospheres, stimulating the reader to establish direct links between worlds in our Solar System and those orbiting other stars. The outcome of a week of intense lectures given by relevant players at the forefront of this field, it constitutes a fundamental reference for advanced undergraduates and Ph.D. students at the early stages of their research career, and its source material is ideal for use in a course setting. Last but not least, with the breadth of its scope this volume is also to be seen as a tribute to the extraordinary diversity, dynamism, and interdisciplinarity of research in the field of extrasolar planetary systems' atmospheres.

Vietri sul Mare, Italy
May 2017

Valerio Bozza
Luigi Mancini
Alessandro Sozzetti

Acknowledgements

The organizing committee of the 2nd Advanced School on Exoplanetary Science would like to thank the Max Planck Institute for Astronomy, the Department of Physics of the University of Salerno, and the Italian National Institute for Astrophysics (INAF) for their financial support. The organizing committee kindly thank the Director of the International Institute for Advanced Scientific Studies (IIASS) in Vietri sul Mare, *Prof. emeritus Ferdinando Mancini*, for hosting the event. Finally, the organizing committee would also like to acknowledge the great help offered by the staff of IIASS, *Mrs. Tina Nappi* and *Mr. Michele Donnarumma*, for the organization of the school.

Contents

Contributors

Jonathan J. Fortney Other Worlds Laboratory (OWL), Department of Astronomy and Astrophysics, University of California, Santa Cruz, CA, USA

Davide Grassi INAF–Institute for Space Astrophysics and Planetology, Rome, Italy

David K. Sing Astrophysics Group, School of Physics, University of Exeter, Exeter, UK

Jonathan Tennyson Department of Physics and Astronomy, University College London, London, UK

Part I
Observational Techniques

Chapter 1
Observational Techniques with Transiting Exoplanetary Atmospheres

David K. Sing

Abstract Transiting exoplanets provide detailed access to their atmospheres, as the planet's signal can be effectively separated from that of its host star. For transiting exoplanets three fundamental atmospheric measurements are possible: transmission spectra—where atmospheric absorption features are detected across an exoplanets limb during transit, emission spectra—where the day-side average emission of the planet is detected during secondary eclipse events, and phase curves—where the spectral emission of the planet is mapped globally following the planet around its orbit. All of these techniques have been well proven to provide detailed characterisation information about planets ranging from super-Earth to Jupiter size. In this chapter, I present the overall background, history and methodology of these measurements. A few of the major science related questions are also discussed, which range from broad questions about planet formation and migration, to detailed atmospheric physics questions about how a planet's atmosphere responds under extreme conditions. I also discuss the analysis methods and light-curve fitting techniques that have been developed to help reach the extreme spectrophotometric accuracies needed, and how to derive reliable error estimates despite limiting systematic errors. As a transmission spectra derived from primary transit is a unique measurement outside of our solar system, I discuss its physical interpretation and the underlying degeneracies associated with the measurement.

1.1 Background and History of Exoplanet Atmosphere Observations

Transiting planets are those that pass directly in front of their parent star as viewed from the Earth. During these events, the planet will block out a proportion of the starlight, which can be detected by time-series photometry. To be viewed in this privileged geometry directly passing in front (or behind) its parent star, transiting planets

D. K. Sing (✉)
Astrophysics Group, School of Physics, University of Exeter, Stocker Road,
Exeter EX4 4QL, UK
e-mail: sing@astro.ex.ac.uk

V. Bozza et al. (eds.), *Astrophysics of Exoplanetary Atmospheres*, Astrophysics and Space Science Library 450, https://doi.org/10.1007/978-3-319-89701-1_1

require a fortuitous orbital alignment with the Earth. As such, transiting exoplanets represent only a small fraction of the total exoplanet population. However, the fundamental properties such as the planetary mass and radius which can be determined (in many cases nearly free from astrophysical assumptions), and the detailed spectroscopic information that can also be measured make transiting exoplanets extremely valuable.

The planet-to-star radius ratio can be very precisely measured during a transit event, as the fractional flux deficit measured from a light curve, $\frac{\Delta f}{f}$, which is proportional to the projected area between the planet and star,

$$\frac{\Delta f}{f} \simeq \left(\frac{R_{pl}}{R_{star}}\right)^2, \tag{1.1}$$

where R_{pl} and R_{star} are the planet and stellar radii respectively. Stellar limb darkening also further modifies the transit light curve shape. As the radii of stars can generally be well determined, the radius of an exoplanet can be measured. Other fundamental properties of the exoplanetary system can also be derived from a transit's light-curve including the planet's inclination, the semi-major axis of the orbit, and the stellar density.

For planetary characterisation, the atmosphere is accessible through the transmission spectrum. When a planet passes in front of its host star, some of the starlight will be filtered through the atmosphere at the planet's terminator and will leave a spectral imprint. Atoms and molecules in the atmosphere will absorb and scatter light at characteristic frequencies, which will make the atmosphere at those wavelengths opaque at higher altitudes. In other words, the atmosphere will be optically thick in slant transit geometry higher in the atmosphere. The exoplanet will therefore have a slightly larger apparent radius at those characteristic wavelengths, which is directly observable via a deeper transit depth (see Fig. 1.1). In a sense, a transmission spectrum is essentially an absorption spectrum, as identification of atomic and molecular species in the planet's atmosphere are identified via absorbed stellar light, though a technically more accurate description is a transit radius spectrum. Exoplanet transmission spectra are now typically constructed by taking time-series spectrophotometry during a transit event, with the spectra divided into many wavelength bins in which the chromatic change in transit depth is measured. At each wavelength bin, a transit light curve must be fit with a model (typically including instrumental systematic effects along with a limb-darkened theoretical transit model) and the planet radius at a particular wavelength, $R_{pl}(\lambda)$, is extracted. A transmission spectrum is very sensitive to the atmospheric composition, which is typically among the first bits of information one learns when a positive signature is identified.

A planet's atmosphere can also be characterised from an emission spectrum. During secondary eclipse, when the planet is seen to pass behind its host star, the flux contribution from the planet drops to zero, isolating the flux from the star. Thus, the light from the planet can be efficiently separated from that of the star.

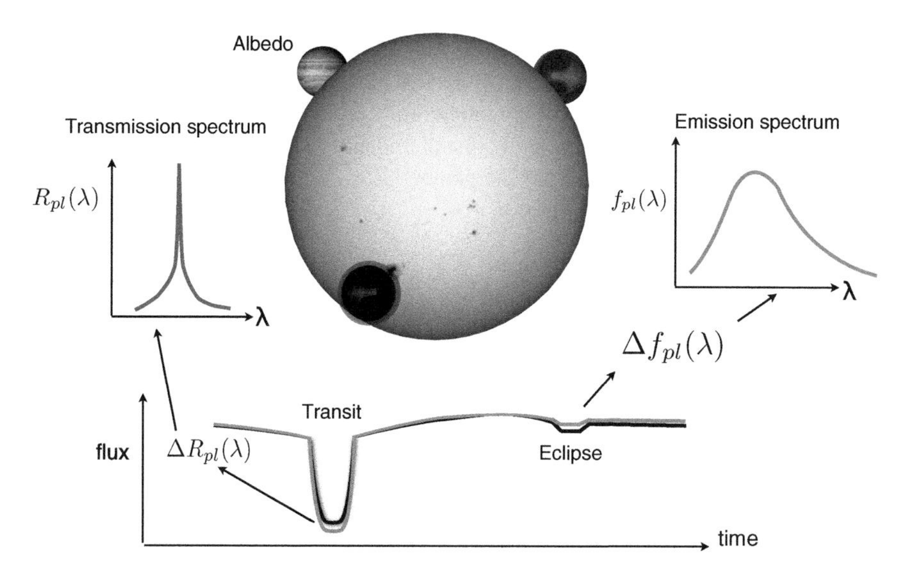

Fig. 1.1 Geometry of an exoplanet transit and eclipse event (top middle). During the exoplanet's orbit, the light curve of the system is monitored, with the transit and eclipse events detected from a drop in flux (bottom). A transmission spectrum is measured by detecting a change in transit depth as a function of wavelength (top left). A thermal emission spectrum is measured by detecting a change in the eclipse depth as a function of wavelength in the infrared (top right). The fraction of reflected light from the planet (the albedo) can be measured by observing the secondary eclipse at optical wavelengths

The eclipse depth measured from the fractional flux deficit at secondary eclipse is directly proportional to the planet-to-star flux ratio, F_{pl}/F_{star} with,

$$\frac{\Delta f}{f} = \frac{F_{pl}}{F_{star}} \times \left(\frac{R_{pl}}{R_{star}}\right)^2 . \tag{1.2}$$

Similarly to a transmission spectra, typically an emission spectra is constructed by taking time series spectrophotometry, and dividing the spectra into many different wavelength bins in which the eclipse depth is measured and the planet-to-star flux ratio is extracted. The amount of flux emitted by a planet at infrared wavelengths depends on its temperature, so this important planetary parameter can be measured with emission spectra in the infrared. Optical secondary eclipses are sensitive to the reflected light and the geometric albedo of an exoplanet.

Finally, when a planet is observed over the course of a full orbit, the flux contribution from the exoplanet will modulate the total star-plus-planet flux as its orbital viewing geometry changes, with atmospheric information obtainable throughout the phase curve. While observing an exoplanetary phase curve does not require a transit or eclipse event, to date most phase curve studies have focused on transiting exoplanets as they typically offer better overall constraints. For instance, the total flux contribution to the phase curve from the star, which dominates the total signal, can only be precisely measured during a secondary eclipse event. From a phase curve, we can measure the day-to-night temperature contrast, which informs us about atmospheric recirculation. Additionally, the abundances of species and atmospheric temperatures can be mapped around the planet.

1.1.1 A Few Early Results

The following sections are by no means a complete summary or census of all exoplanet observations relating to atmospheric characterisation that have been obtained, nor is it meant to be. Further results can be found in several review articles (e.g. Seager and Deming 2010; Burrows 2014; Bailey 2014; Crossfield 2015; Deming and Seager 2017). The intention is to highlight a few representative works to give the reader a broad introductory overview of the types of atmospheric measurements that have and can be made regarding transiting exoplanets, and a flavour of the sort of scientific investigations that result.

1.1.2 Transit Observations

As recently as 2006 there were fewer than 10 transiting exoplanets known, and the atmospheric characterisation of these planets were largely limited to a few select cases. Among the first characterised planets was HD 209458b, a exoplanet first found using the radial velocity technique (Mazeh et al. 2000) which was the first planet discovered to also transit by Charbonneau et al. (2000) and Henry et al. (2000). Having initially been discovered by the radial velocity technique helped make HD 209458b particularly well suited for followup atmospheric characterisation. The radial velocity method requires bright target host stars in order to obtain sufficient signal-to-noise at high spectral resolution, which has historically limited radial velocity discoveries on moderately sized telescopes to V-magnitudes typically brighter than about 10. Transiting around such bright host stars opens up the possibility to perform very high precision photometry, as sufficiently large numbers of photons can be gathered during the short $\sim$hour long transit durations. Shortly after the discovery of transits, the Hubble Space Telescope (HST) observed HD 209458b during four transits with the Space Telescope Imaging Spectrograph (STIS). The resulting transit light curve (Brown et al. 2001) was unprecedented in quality with 110 parts-per-million photometric accuracies with an 80 s cadence. The high quality demonstrated not only the feasibility to perform atmospheric studies but also to detect Earth-sized exoplanets in transit around Sun-like stars (the NASA Kepler mission was selected 12 months later). With the same STIS data, Charbonneau et al. (2002) made the very first detection of an exoplanet atmosphere by observing excess absorption in the Na doublet (See Fig. 1.2). Alkali metal absorption had been previously predicted to be present

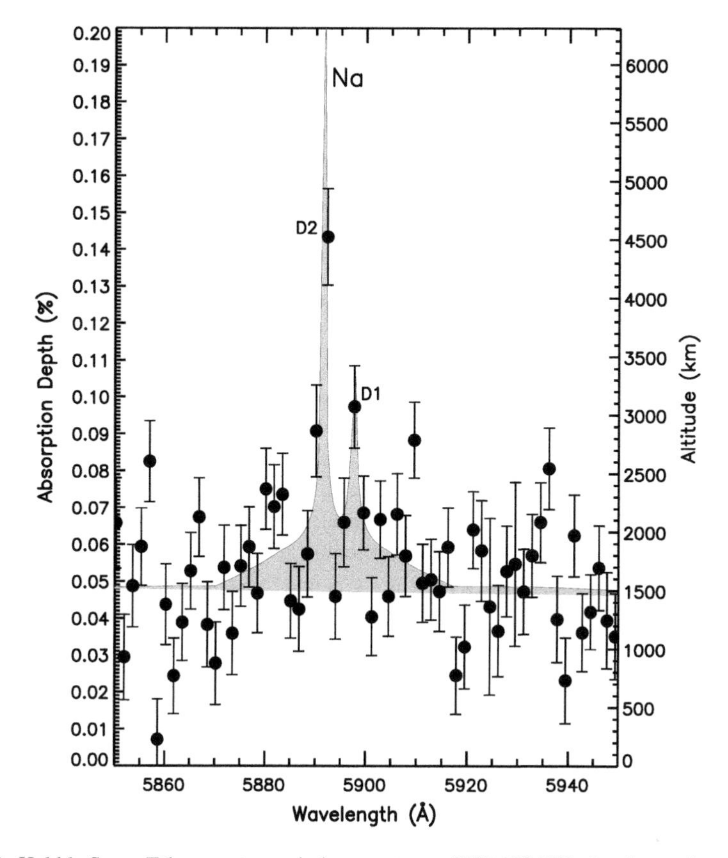

Fig. 1.2 Hubble Space Telescope transmission spectrum of HD 209458b showing sodium absorption, which was first detected by Charbonneau et al. (2002). Shown is the transmission spectrum adapted from Sing et al. (2008), which resolves the sodium doublet. The relative transit absorption depth and relative altitude in the planetary atmosphere are both indicated

in the atmosphere by Seager and Sasselov (2000), as the high temperature allows atomic sodium to exist in the gas phase.

The HD 209458b sodium detection was a good early indication of the high photometric precisions which would be needed to regularly make exoplanet atmosphere detections, as the observed sodium signal measured by Charbonneau et al. (2002) represented a deeper transit by just 232 parts-per-million (ppm) relative to the adjacent wavelengths and precisions of 57 ppm were needed to secure a 4-σ confident detection.

As the transit technique is reliant upon the starlight to make a measurement, and not light from the planet itself, a transmission spectrum can be studied across a much broader wavelength range than traditionally possible when observing only emitted radiation. Shortly after the sodium detection, HST STIS followed HD 209458b up again but this time in the far-ultraviolet (FUV), targeting the H I Ly-α line. An extended exosphere of H I was expected around the exoplanet, which could in principle be detected against the stellar chromospheric Ly-alpha emission during a transit event. However, the resulting transit depth as observed in H I (15%, Vidal-Madjar et al. 2003) was far in excess of the planet itself or even its Roche lobe, indicating hydrogen atoms are vigorously escaping from the planet. Oxygen and carbon were also detected in the exosphere by STIS shortly thereafter (Vidal-Madjar et al. 2004). With the second transiting planet discovered in 2003 (Konacki et al. 2003), for a short time, the number of atmospheric detections made via the transit method outpaced the number of transiting planets. Even with thousands of transiting planets known today, HD 209458b still resides as one of the very best theoretical targets for atmospheric characterisation.

Transmission spectral observations from the ground also proved feasible, first at high resolution. Redfield et al. (2008) observed HD 189733b across optical wavelengths with the Hubby-Eberly Telescope at high resolution (spectral resolution, R $\sim$ 60,000). Significant absorption was detected in-transit in the Na D lines, indicating deeper transit depths by 672 ± 207 ppm. Similarly, Snellen et al. (2008) analysed optical HD 209458b transit data from the Subaru telescope at high resolution ($R \sim$ 45,000). A significant non-linearity in the CCD had to be corrected, and once applied significant Na absorption was also detected (1350 ± 170 ppm). The Na absorption from HD 209458b first observed with Hubble was not only confirmed, but the Na line profile itself also matched well between the observations (Charbonneau et al. 2002; Sing et al. 2008; Snellen et al. 2008). Historically, such agreement between observations has not always been case, but these early results did prove to place exoplanet atmospheric observations on a solid foundation.

The shape of absorption line profiles in transmission spectra can be used to probe different altitudes of exoplanet atmospheres. The wings of a line have a lower optical depth, and probe lower, cooler parts of the atmosphere, while the extended core of the line probes out to much higher altitudes. Because the atomic Na D resonance lines are very strong and have little other absorbers obscuring the view in their wavelength region, they are ideal for probing a wide altitude-pressure range. Their line profiles have been well-measured from HST and the ground for HD 189733b and HD 209458b (see also Huitson et al. 2012; Jensen et al. 2011). Thermosphere layers were detected in both planets (e.g. Vidal-Madjar et al. 2011 for HD 209458b and Huitson et al. 2012 for HD 189733b). The thermosphere is an extended region above the typical lower layers of the atmosphere (troposphere and stratosphere), but below the exosphere, where UV radiation is absorbed and temperatures rise as a function of height. The cores of the Na lines extend to higher altitudes in the presence of a thermosphere, as the hotter thermospheric temperatures increase the pressure scale height and puff up the atmosphere—leading to larger transmission spectral features. For instance, HARPS observations of HD 189733b have resolved the Na lines up to an altitude of 12,700 km (Wyttenbach et al. 2015). With high enough signal-to-noise (S/N), a change in pressure scale height at different altitudes can be directly detected, which then informs us about the temperature change in the atmosphere between the upper and lower layers (see Sect. 1.4).

1.1.3 Eclipse Observations

In August of 2004 the HST STIS instrument failed, preventing further observations until a repair could be made. However, a year earlier the Spitzer Space Telescope was launched, and while it was not designed for exoplanet transit observations, the first secondary eclipse measurements were achieved shortly after in 2005 by Deming et al. (2005) for HD 209458b, while Charbonneau et al. (2005) targeted TrES-1b. As Spitzer is an infrared telescope, it is sensitive to the longer wavelengths needed to probe a hot exoplanet closer to its ~1000 K black-body emission peak. Additionally, at these longer wavelengths the parent star is significantly fainter, which further increases the contrast. Both exoplanets were observed to have secondary eclipse depths near 0.25% and from these single eclipse measurements, brightness temperatures (which assumes the planet emits as a black body) could be derived. The timing of the secondary eclipse measurements placed informative constraints on the orbital eccentricity. With these results, Spitzer in effect took the reins over a period of a few years as the leading (and virtually only) exoplanet characterising instrument. The repair of HST and installation of the WFC3 instrument in 2009 also enabled high-precision secondary eclipse measurements (e.g. Fig. 1.3).

With multiple secondary eclipse measurements at different wavelengths, Spitzer was able to build up the first rough broadband emission spectra, which could be compared to theoretical atmospheric models. An early result came from Knutson et al. (2008) who used Spitzer to build up a spectrum from five photometric channels between 3 and 30 microns for HD 209458b. These broadband spectra are very low resolution ($R \sim 5-10$), thus no specific molecular features can be resolved, though in principle very large-scale differences between atmospheric models are detectable. The HD 209458b measurements indicated the presence of a stratosphere where the temperature is seen to rise (rather than fall) at higher altitudes. Rising temperatures create conditions where hot atmospheric gas lies above cooler gas, which gives rise to spectral emission features. If a planetary atmosphere has a temperature-pressure (T-P) profile which only cools off with higher altitudes, only spectral absorption signatures from the overlying cooler gas could be observed. For HD 209458b, Knutson et al. (2008) found the wavelengths between 4 and 10 microns were significantly higher than would be expected from non-inverted T-P profiles which indicated H_2O was in emission (also see Burrows et al. 2007). While these specific measurements did not generally hold up to further scrutiny (see Sect. 1.2.2), they were a strong motivation into further sophisticated theoretical investigations of the atmospheric dynamics and chemistry of highly-irradiated gas giant planets, and the first steps towards constraining such models with eclipse observations.

Atmospheric windows in Earth's own atmosphere also permit exoplanet transit and eclipse observations from the ground. However, the challenges are formidable as precision photometry (~100 ppm) must be preformed for several hours against strongly changing weather conditions and instrument instabilities. Furthermore, for eclipse observations the largest signals occur at longer wavelengths which are generally inaccessible due to telluric H_2O absorption. Thus, exoplanet eclipse observations

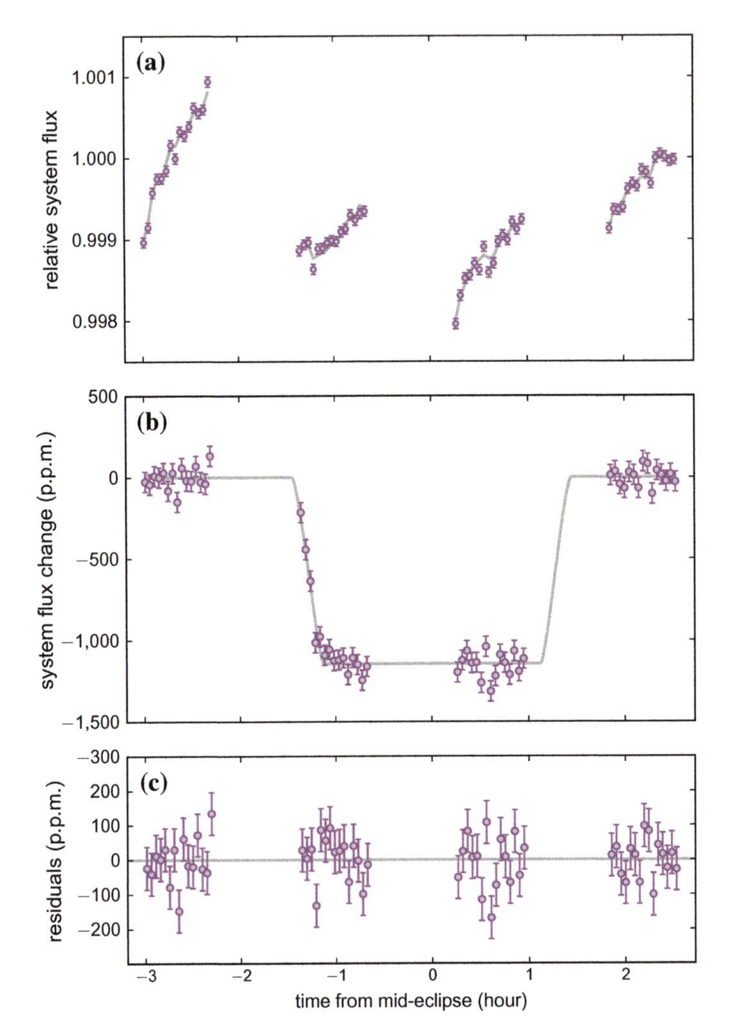

Fig. 1.3 Example of a secondary eclipse observation of WASP-121b from the Hubble Space Telescope WFC3 (from Evans et al. 2017). Plotted **a** is the raw normalized flux with photon noise error bars and the best-fit eclipse and instrument systematic trend model, **b** the relative change in system flux after correcting for instrument systematics, and **c** the residuals between the data and best-fit model showing precisions of 64 parts-per-millon

have generally been performed in the z', J, H, and K atmospheric windows. Theoretical predictions early on (López-Morales and Seager 2007) predicted the hottest planets (~2500 K) would likely have significant and detectable thermal emission in the red-optical z' band, and shortly after Sing and López-Morales (2009) detected the thermal emission of Ogle-Tr-56b while at the same time de Mooij and Snellen (2009) detected the secondary eclipse of TrES-3b in the K band with the WHT telescope. These results directly demonstrated a very large number of exoplanet eclipses would eventually be observable, given that Ogle-Tr-56b orbits a very distant 16th magnitude star and the TrES-3b result was made with a modestly sized 4-meter telescope. Croll et al. (2011) observed eclipses of WASP-12b with the CFHT telescope at J, H, and K band which further demonstrated high-precision measurements could be obtained from the ground. Thus, exoplanet atmospheric characterisation from eclipses were no longer confined to flagship space telescopes, and transit and eclipse observations became demonstrably accessible for a wide range of targets using a wide variety of instrumentation.

1.1.4 Phase Curve Observations

In general, the most difficult atmospheric measurement to make with a transiting exoplanet is that of the phase curve. The difficulty stems from the requirement to maintain high photometric precisions of order 100 ppm (which, as with transits and eclipses, are still necessary) over long timespans. Typically, measurements last on the order of the orbital period of the planet, which can be a day or longer—compared to transit/eclipse events which last a few hours. Furthermore, taking into account the very large telescope times needed, for practical purposes phase curve observations thus far have been focused on the shortest period planets with periods of 2 days or shorter.

While early non-continuous observational attempts were made (Harrington et al. 2006), a notable phase curve observation was made by Knutson et al. (2007b), who observed HD 189733b with the Spitzer Space Telescope at 8 microns during half an orbital period covering a transit and eclipse. The measured phase curve amplitude indicated a modest day-night circulation, which results in more modest phase curve amplitudes. In addition, the brightest portion of the phase curve was observed just before secondary eclipse. For a tidally locked planet, one may expect the hottest and thermally brightest point on the planet to occur at the sub-stellar point. However, strong atmospheric winds (as predicted by Showman and Guillot 2002) have the effect of advecting the heat and hottest part of the planet westward of the substellar point. Longitudes westward of the sub-stellar point are maximally viewable just before the secondary eclipse event itself, leading to an observed light curve where the maximum flux is found before an eclipse.

An important aspect of phase curve observations done on transiting planets is that a transmission spectrum and an eclipse spectrum can also be derived from the same dataset, making it a particularly constraining measurement. An example can be seen in Fig. 1.4 where a spectroscopic phase curve of WASP-43 was observed by Stevenson et al. (2014, 2017) with HST and Spitzer. In the phase-curve spectra, H_2O features were mapped around the planet, and the emission spectrum showed strong absorption features.

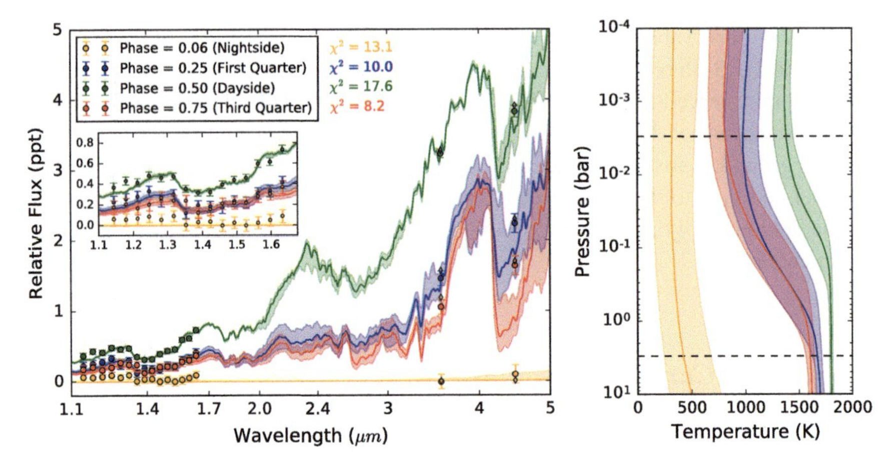

Fig. 1.4 Hubble Space Telescope WFC3 and Spitzer Space Telescope phase curve spectrum of WASP-43b at four different orbital phases (left) and the corresponding retrieved temperature-pressure profiles (right). Each set of colored circles depict measurements from the phase curve observations, and the colored curves with shaded regions represent the median models with 1σ uncertainties. The inset magnifies the WFC3 spectra. H_2O absorption is readily apparent in the spectra at 1.4 microns where a significant drop in the planet-to-star flux ratio is apparent. From Stevenson et al. (2017)

1.1.5 Accessible Transmission Spectra Exoplanets

In 2006, there were 158 known exoplanets, with the majority of the population found from the radial velocity technique. Moreover, only a very small handful of exoplanets had their atmospheres regularly detected, most notably HD 209458b and HD 189733b. Not all exoplanets are ideal atmospheric targets. The more massive exoplanets have smaller atmospheric scale heights due to the high surface gravity, and will have smaller transmission spectral signals. For secondary eclipses, cooler exoplanets will have low thermal fluxes in the optical and near-IR, so will be more challenging to detect at those wavelengths.

For an exoplanet transmission spectrum, a good indication of the expected signal can be estimated by calculating the contrast in area between the annular region of the atmosphere observed during transit and that of the star. The characteristic length scale of the atmosphere is given by the pressure scale height,

$$H = \frac{k_B T}{\mu g} \tag{1.3}$$

where k_B is the Boltzmann constant, T is the temperature of the atmosphere, μ is the mean mass of atmospheric particles, and g is the surface gravity. For giant exoplanets, the composition of the atmosphere can be assumed to be dominated by a H/He mixture of near-solar composition, which gives $\mu = 2.3 \times u$ where u is the unified atomic mass unit (Lecavelier des Etangs et al. 2008). For transiting exoplanets, the surface gravity is well known for a large majority of the prime transmission spectral targets, as historically the radial velocity method has been used to confirm the planetary nature of a transiting object. Those exoplanet systems too faint to be detected via radial velocity are also often too faint to perform detailed atmospheric characterisation. Moreover, knowing the mass of an exoplanet for transmission spectroscopy is vital, as large degeneracies will be present if the mass is unknown which will limit the usefulness in constraining the atmospheric properties.

A good way to estimate the atmospheric temperature is to use the equilibrium temperature value, T_{eq}. Assuming zero albedo and complete redistribution of heat around the planet, T_{eq} can be calculated using,

$$T_{eq} = (1/4)^{1/4} T_{eff} \sqrt{\frac{R_{star}}{a}} \tag{1.4}$$

where a is the semi-major axis of the planet and T_{eff} is the stellar effective temperature (Cowan and Agol 2011). The temperatures derived thus far from transmission spectra have often been well within these equilibrium values. For example, HD 189733b, HAT-P-12b and WASP-6b have temperatures derived from their transmission spectra of 1340 $\pm$ 150 K, 1010 $\pm$ 80 K and 973 $\pm$ 144 K, respectively which compare very favourably to their T_{eq} values of 1200, 960, and 1150 K, respectively (Sing et al. 2016). With the scale height estimated, it is straightforward to approximate the absorption signal, A, of the annular area of one atmospheric scale height H during transit, as

$$A = \frac{(R_{pl} + H)^2}{R_{star}^2} - \left(\frac{R_{pl}}{R_{star}}\right)^2, \tag{1.5}$$

which can be further simplified assuming $H << R_{pl}$ to,

$$A = \frac{2R_{pl}H}{R_{star}^2}. \tag{1.6}$$

Transmission spectral signals are typically on the order of 1 to $\sim 5H$ in size, thus if the transit depth can be measured to about $1H$ in precision with sufficient spectral resolution, detectable spectral features would begin to appear. Plotting A against the magnitude of the host stars is a good proxy to compare the relative signal-to-noise of different exoplanets. While other factors such as cloud cover will ultimately determine if atmospheric features will be present or not, all other factors being equal, it is often a good guide to prioritise exoplanets with the largest expected signal-to-noise values. From Fig. 1.5, the prominence of HD 189733b and HD 209458b become apparent. Both exoplanets orbit much brighter stars (V $\sim$ 7.7) than the bulk of the known transiting planets, which dramatically improves the photon noise limits. The "puffiest" planets, like WASP-17b may orbit a much fainter star (V = 11.6) but the

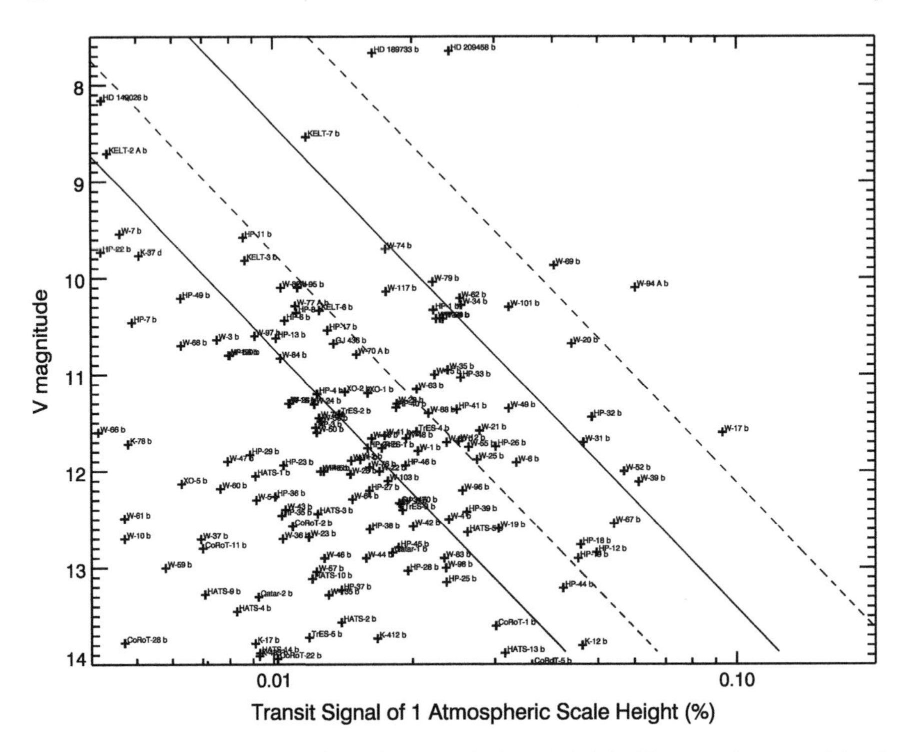

Fig. 1.5 Transmission spectral signal of 1 atmospheric scale height. Planets to the upper-right of the figure are easier to observe. The lines indicate approximately constant S/N values. The exoplanet data was compiled from the exoplanets.org database. The WASP and HAT-P planetary names have been abbreviated

expected atmospheric transit signal is large ($A \sim 0.09\%$) making it a comparable target to HD 189733b in terms of expected S/N. Perhaps 100 or more transiting exoplanets are now accessible with today's instruments. In practice, other practical considerations are necessary to take into account. For instance, ground-based multi-object spectroscopy requires reference stars to perform differential spectrophotometric measurements. However, few, if any suitable reference stars would likely be available for observing a 7th magnitude target with a typical 4–8 m class telescope. Such instruments have typical fields of view of around 10 arc minutes, and it is unlikely that more than one bright (and hence nearby) star would be close to each other in the sky.

1.1.6 Exoplanets with Accessible Secondary Eclipses

Not all exoplanets are favourable for secondary eclipse measurements. The expected eclipse depths can be estimated in a similar exercise as for transmission spectra.

However, the results will be much more dependent on the observed wavelength, because we measure the relative flux contrast between the planet and the star, which radiate at very different blackbody temperatures. The dayside flux from the planet and secondary eclipse depth, $\Delta f_{day}/f$ can be estimated from

$$\frac{\Delta f_{day}}{f} = p_\lambda \left(\frac{R_{pl}}{a}\right)^2 + \frac{B_\lambda(T_{day})}{B_\lambda(T_{eff})} \times \left(\frac{R_{pl}}{R_{star}}\right)^2, \tag{1.7}$$

which takes into account a reflection component with the wavelength dependant albedo p_λ, and a thermal component $B_\lambda$which is approximated here assuming the planet radiates as black body with temperature T_{day}, and the star also radiates as a black body with temperature T_{eff} (Haswell 2010). Similarly to Fig. 1.5, the expected secondary eclipse depth at a given wavelength can then be plotted against the host star magnitude, at that wavelength range, to assess the relative observability of different targets. Of course, in reality stars are not blackbodies, and neither are planetary atmospheres so there can be significant deviations from such simple estimations. Nevertheless, such plots as shown in Fig. 1.6, can help illustrate the relative potential signal sizes between planets, all else being equal.

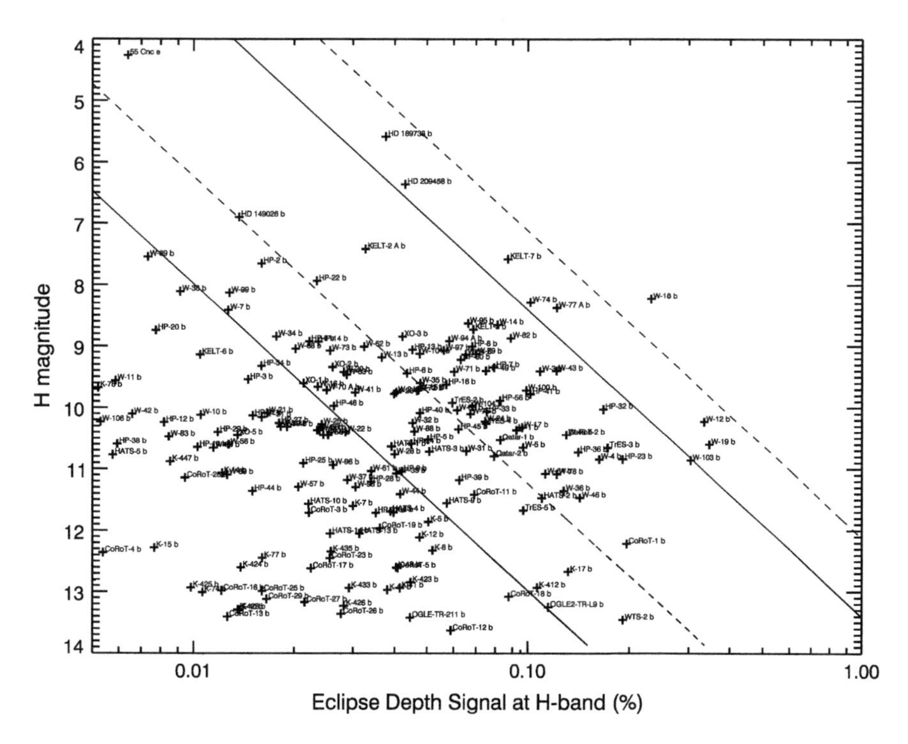

Fig. 1.6 Expected secondary eclipse signals at H-band. Planets to the upper-right of the figure are easier to observe. The lines indicate approximate constant S/N values. The exoplanet data was compiled from the exoplanets.org database. The WASP and HAT-P names have been abbreviated

1.1.7 Accessing the Atmospheres of Small Exoplanets

Since the transit technique is predominately limited by the flux of the host star, it can be used across a much broader wavelength range than secondary eclipse measurements. Transits have been measured in the UV (e.g. Vidal-Madjar et al. 2003) through to the far-infrared (Richardson et al. 2006). Additionally, much smaller exoplanets are currently accessible with transit spectroscopy than with other techniques. There is strong interest to push characterisation down to smaller, cooler planets and so toward potentially life bearing worlds. The task is difficult, as small planets and cooler temperatures result in much smaller transit atmospheric signatures (see Fig. 1.7). To overcome this difficulty, currently there are two strategies being pursued to meet the near-term goal of detecting atmospheric features and viably searching for biomarker signatures within the atmospheres of extrasolar planets; transit spectroscopy of exoplanets around small M-dwarfs (the 'M-dwarf opportunity') or around very bright stars (the 'bright-star opportunity'). These opportunities have driven dedicated transiting M-dwarf searches, such as MEarth (Charbonneau et al. 2009), as well as dedicated space missions such as TESS and PLATO which will search the brightest stars for transits.

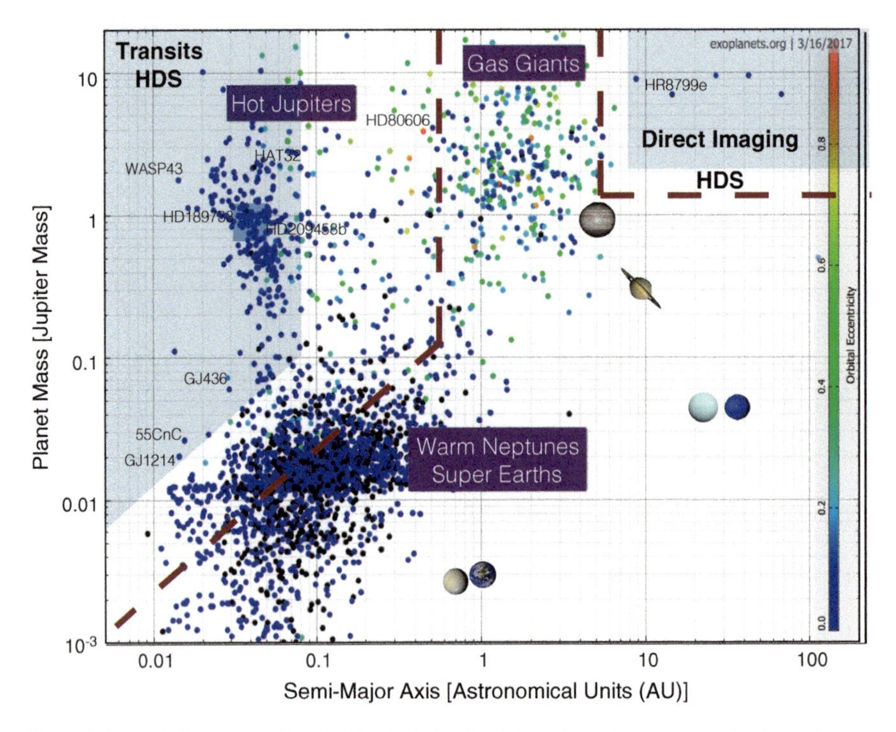

Fig. 1.7 Planet Mass versus Semi-Major Axis for the detected exoplanets, compiled from the exoplanets.org database. The broad exoplanet types are labeled and solar system planets also indicated. The approximate current sensitivity of atmospheric studies is indicated (dark shaded region) and near-term expected improvements (red dashed lines)

While searching for signatures of habitability will be an important long-term exoplanet goal, it must be put into perspective as the very wide and diverse group warm-Neptunes, super-Earths and small terrestrial exoplanets represent a completely uncharted parameter space. Developing a comprehensive theory to explain the atmospheres of these planets more generally, and therefore put any atmospheric detection within a wider context will represent an enormous challenge.

The discovery of the super-Earth GJ1214b orbiting an M-dwarf (Charbonneau et al. 2009) provided the first atmospheric glimpses into small exoplanets (Bean et al. 2010). GJ1214b orbits a small, M4.5 dwarf star, which means it has a large transit depth and atmospheric transmission signal. Both of these observable quantities scale inversely with the radius of the star squared, so the signals of exoplanets orbiting smaller stars are greatly enhanced. For GJ1214b, the star has a radius of $R_{star} = 0.2R_{\odot}$ making an exoplanet signal $(1/0.2)^2 = 25\times$ higher than if the same planet orbited a sun-like star. The discovery of TRAPPIST-1b, c, d, e, f, g (Gillon et al. 2017) has pushed the M-dwarf opportunity to even smaller planets and stars, with the $0.114\ R_{\odot}$ sized M8 star providing transmission spectral signals $77\times$ higher than if the same planet orbited a sun-like star. To this end, even the atmosphere of Trappist-1f, which is an Earth-sized exoplanet orbiting within the habitable zone and is expected to be rocky, may potentially be detected in the near-future. Even with a large signal boost from a small star, the atmospheric features are still expected to be small and challenging to detect. Current facilities such as HST have allowed H-dominated atmospheres to be ruled out on several of the TRAPPIST-1 planets (de Wit et al. 2016).

On the bright end, the discovery of a planet transiting around HD 97658b (Dragomir et al. 2013) permitted atmospheric investigations of a super-Earth around a much more massive K1V star. In both the case of HD 97658b and GJ1214b, the exoplanet's atmospheres proved to be largely consistent with heavy cloud-cover and no spectral features were detected (Knutson et al. 2014; Kreidberg et al. 2014b).

1.2 Exoplanet Atmosphere Science Topics

In the following sections, I have highlighted a few current science topics which transiting exoplanet atmosphere observations can (or will hopefully) address. While there are a very wide and diverse range of science topics, many fall under a few basic categories which are briefly described below.

1.2.1 Planet Formation

Gas giant exoplanets are predominantly composed of a H/He mixture, which was accreted during the planets formation from the protoplanetary disk. As such, the gas is primordial in nature so may be expected to contain records of the formation conditions in which the planet formed. As transiting exoplanets are amenable to spectroscopic studies, one may expect then to probe what would be essentially the primordial gas and gain insights into the planet formation process.

There are two widely considered theories for how gas giant planets form: gravitational instability and core accretion. Gravitational instability is said to occur when the protoplanetary disk rapidly cools and collapses into planetary-mass fragments (Boss 1997). Planets formed via this mechanism would have the same bulk compositions as their local protoplanetary disk material and their host stars. Alternatively, in the core-accretion model, giant planets form in a multi-step process: first, sticky collisions of planetesimals lead to the formation of protoplanetary cores; then, once the cores reach a threshold mass they accrete nearby gas in a runaway fashion (Pollack et al. 1996). Population synthesis models from Mordasini et al. (2012) and Fortney et al. (2013) suggest that in the core accretion paradigm, as the mass of a planet decreases, its atmospheric metallicity increases. This is because lower mass planets would be unable to accrete substantial gas envelopes, and thus would be more susceptible to pollution by in falling, higher metallicity planetesimals. The giant solar system planets agree with the latter scenario (see Fig. 1.8), as the metallicities derived from the methane abundance of Jupiter (from the Galileo probe: Wong et al. 2004), Saturn, Neptune, and Uranus (from infrared spectroscopy: (Fletcher et al. 2009; Karkoschka and Tomasko 2011; Sromovsky et al. 2011), respectively) show decreasing metal enhancement with increasing planet mass. Via these two theories, exoplanetary atmospheres will exhibit different atmospheric properties which can be measured from transmission and emission spectroscopy. Gravitational instability theory suggests that planets will have the same atmospheric metallicity as the central star, while in core accretion theory lower mass planets will have higher atmospheric metallicity.

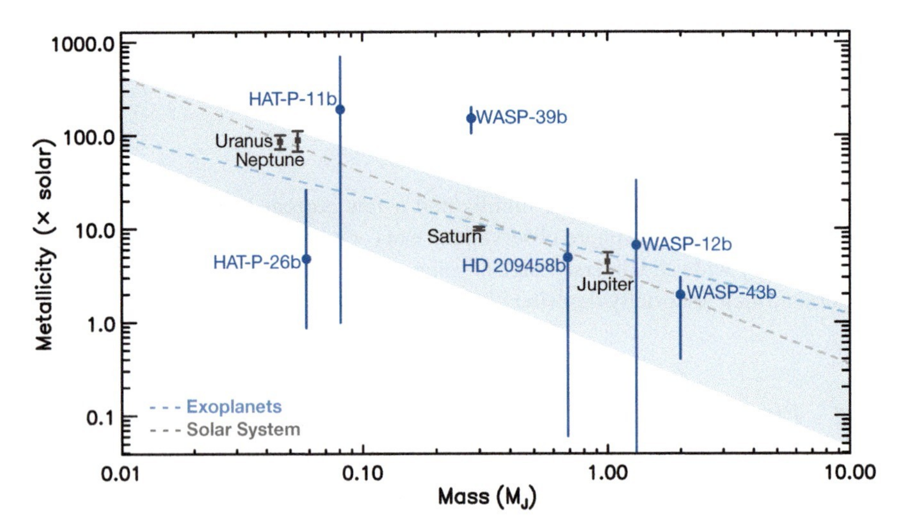

Fig. 1.8 Mass-Metallicity trend for solar system planets and exoplanets, adapted from Wakeford et al. (2017) to include measurements of WASP-39b (Wakeford et al. 2018), and HD 209458b (Line et al. 2016)

Gas-giant exoplanets are widely expected to have retained the bulk of their primordial atmospheres, and measuring the atmospheric abundances across a wide range of planet masses should provide insight into formation mechanisms. Kreidberg et al. (2014a) found evidence the $2M_J$ hot Jupiter WASP-43b follows the same inverse mass-metallicity relationship as the solar system planets. However, the Neptune-mass exoplanet HAT-P-26b has a measured water abundance at just $4\times$ solar (Wakeford et al. 2017) which is below the trend and suggests a different formation and/or evolutionary processes. HAT-P-26b is consistent with recent envelope accretion models (Lee and Chiang 2016), which argue that most hot Neptunes accrete their envelopes in situ shortly before their disks dissipate. In both studies, the retrieved atmospheric water abundance was used as a proxy for the overall planet's metallicity but the abundant carbon-bearing molecules will need to be measured before stringent metallicity constraints are available.

The abundance ratio between carbon-bearing molecules and oxygen-bearing molecules (C/O) is also expected to play a key role in constraining planet formation and migration mechanisms (Madhusudhan et al. 2014). The C/O ratio contains vital information such as the physical properties of the accretion disk in which the planet formed, and a planets location within that disk. This is especially pertinent for hot Jupiters, which migrated to short orbital periods though may have initially formed beyond the snow line (e.g. Öberg et al. 2011). Carbon-rich atmospheres would point to scenarios where hot Jupiters were initially located beyond the snow line and accreted primarily carbon-rich gas, while O-rich atmospheres would point instead to accretion of primarily oxygen-rich solid material (Espinoza et al. 2017).

1.2.2 Atmospheric Physics

Transiting exoplanets represent a new novel laboratory in which to test our models of atmospheric physics. The temperatures, gravities, and chemical compositions occupy a very wide and new range of physical conditions, making observations of these planets capable of giving new broad physical insights into how planetary atmospheres operate. Current theories of hot gaseous planets contain many open questions about their atmospheric characteristics (temperature, clouds, energy budget, atmospheric escape), their chemical abundances, and how they formed and evolved. All these questions are intertwined, and by observing, characterising, and comparing many exoplanets across a broad parameter space, progress on answering some of those questions can be made.

Transiting planets orbit close to their host stars, making them tidally locked and highly irradiated. Those factors affect the planet's vertical and horizontal (day-to-night) temperature structure, and induce photochemical processes in their atmospheres, which do not occur in the most related astrophysical objects, isolated brown dwarfs (BD). Therefore, hot exoplanets are completely new objects with a set of physical processes that are uniquely challenging to theoretically model.

From the first observations of secondary eclipse spectra, the thermal structure of hot Jupiters has been an active area of theoretical investigations and observational efforts. One of the first planets characterized through secondary eclipse measurements showed evidence for a thermal inversion and hot stratospheric layer (Knutson et al. 2008). A hot stratosphere is caused by strong optical absorbers, which absorb stellar radiation at altitudes higher than they thermally radiate energy, which heats the upper atmosphere and causes a stratospheric layer (Hubeny et al. 2003; Burrows et al. 2007; Fortney et al. 2008). On the Earth, UV absorption by ozone creates a stratospheric layer and most solar system planets including Jupiter and Saturn have stratospheres (Gillett et al. 1969; Wallace et al. 1974; Ridgway 1974).

The presence or absence of a stratosphere is expected to change the global energy budget and atmospheric circulation and dynamics of the planet, making their presence and theoretical understanding an important aspect of their overall atmospheric makeup. In highly irradiated gas giant exoplanets that lack a strong optical absorber, the incident stellar irradiation is absorbed deep in the atmosphere, near pressures of 1 bar (see Burrows et al. 2008). This pressure is close to the near-IR emission photosphere, resulting in a monotonically decreasing temperature profiles and a lack of a stratosphere. At these pressures, the expected wind speeds in a hot Jupiter will be able to efficiently redistribute heat around the entire planet, leading to modest day/night temperature contrasts. With a strong optical absorber at high altitudes, the local gas is radiatively heated by the incident stellar flux, creating a stratosphere. In addition, the winds at these lower pressures (higher altitudes) are not able to efficiently redistribute the energy at their near-IR photospheres, creating a very strong day-night temperature contrast. With a stratosphere, the hottest part of the planet becomes located at the highly irradiated sub-stellar point, while the atmosphere becomes cooler towards the limb.

Several candidates for strong optical absorbers at altitude were proposed (Hubeny et al. 2003; Burrows et al. 2007; Fortney et al. 2008) with TiO/VO being the currently leading candidates. Fortney et al. (2008) highlighted the importance of gaseous TiO and VO to the optical opacities of highly irradiated hot-Jupiters, proposing two classes analogous to M and L-type dwarfs. In this scenario, hot-Jupiters warm enough to still have gaseous TiO and VO were dubbed "pM Class" planets. This class contains temperature inversions, and appears "anomalously" bright in the mid-infrared at secondary eclipse, as the stellar incident flux is absorbed high in the atmosphere and emitted as thermal flux at near-IR wavelengths. Theoretical models predicting the transmission spectra of pM class of planets would be dominated in the optical by TiO opacity (Fortney et al. 2008). The optical transmission spectra of cooler pL Class planets (lacking TiO) are thought to be dominated by neutral atomic Na and K absorption, and lack hot stratospheres.

Follow-up studies of HD 209458b with more advanced data analysis techniques did not support the presence of a stratosphere (Diamond-Lowe et al. 2014; Evans et al. 2015). In addition, despite many dozens of exoplanets searched for signatures of stratospheres with Spitzer, no definitive detections were made and confirmed. For a while it seemed TiO/VO may not be present in hot-Jupiter atmospheres. Spiegel et al. (2009) argued that vanadium oxide was not likely to fulfil this role due to low abundances, and that the previously favoured titanium oxide would require unusually high levels of macroscopic mixing to remain in the upper atmosphere.

A decade later, the topic of hot Jupiter stratospheres is still a hot topic as a thermally inverted spectral signature was observed for WASP-121b with H_2O seen in emission (see Fig. 1.9 and Evans et al. 2017), and evidence for a stratosphere and TiO seen in WASP-33b as well (Haynes et al. 2015; Nugroho et al. 2017). Compared to earlier studies, WASP-121b and WASP-33b are much hotter ($T_{eq} > 2500$ K) than the planets probed earlier, which could indicate much hotter temperatures are required than earlier theoretical studies indicated (Fortney et al. 2008). However, it remains unclear why some planets would have stratosphere layers, while other seemingly similar very hot planets do not.

1.2.3 Clouds and Hazes

Cloud and haze aerosols are ubiquitous for the planets with significant atmospheres within our own solar system. For hot exoplanets currently amenable to transit characterization, clouds and hazes have also been found. The atmospheric temperatures of the hot Jupiters are close to the condensation temperatures of several abundant components, including silicates and iron. The formation of condensate clouds and hazes is a natural outcome of chemistry in much the same way H, C, and O combines to form H_2O, CO, and CH_4. The possible presence of such condensation clouds was considered early on (Seager and Sasselov 2000). Cloud and haze aerosols can form via condensation chemistry, or alternatively the aerosols may be photochemical in nature (e.g. Helling et al. 2008; Marley et al. 2013). Silicate and high-temperature cloud condensates are expected to dominate the hotter atmospheres, while in cooler atmospheres sulphur-bearing compounds are expected (Visscher et al. 2010; Morley et al. 2012; Wakeford and Sing 2015). The presence or absence of clouds and hazes have strong implications on all aspects of a planet's atmosphere including the radiation transport, chemistry, total energy budget, and advection (Marley et al. 2013). As such, the presence of clouds and capacity to model them is currently a major uncertainty and limitation in our ability to interpret exoplanet spectra and retrieve accurate molecular abundances.

According to models, condensates would weaken spectral features, or mask some of them, depending on the height of the cloud deck (Marley et al. 1999; Sudarsky et al. 2003; Fortney 2005). In transmission spectra, a grey cloud (for example) can mask all absorption features below the altitude of the cloud deck, and is an explanation for the muted water feature of HD 209458b (Deming et al. 2013) and the absence of features on the super-Earth GJ1214b (Kreidberg et al. 2014b). The HST transmission spectrum of HD 189733b was found to contain a high altitude scattering haze (Pont et al. 2008), which has since been confirmed by multiple follow-up HST measurements (Sing et al. 2011; Huitson et al. 2012; Gibson et al. 2012b; Pont et al. 2013). For HD 189733b, the haze covers the entire optical regime, with a Rayleigh scattering slope masking all but the strong Na I line cores and likely extends into the

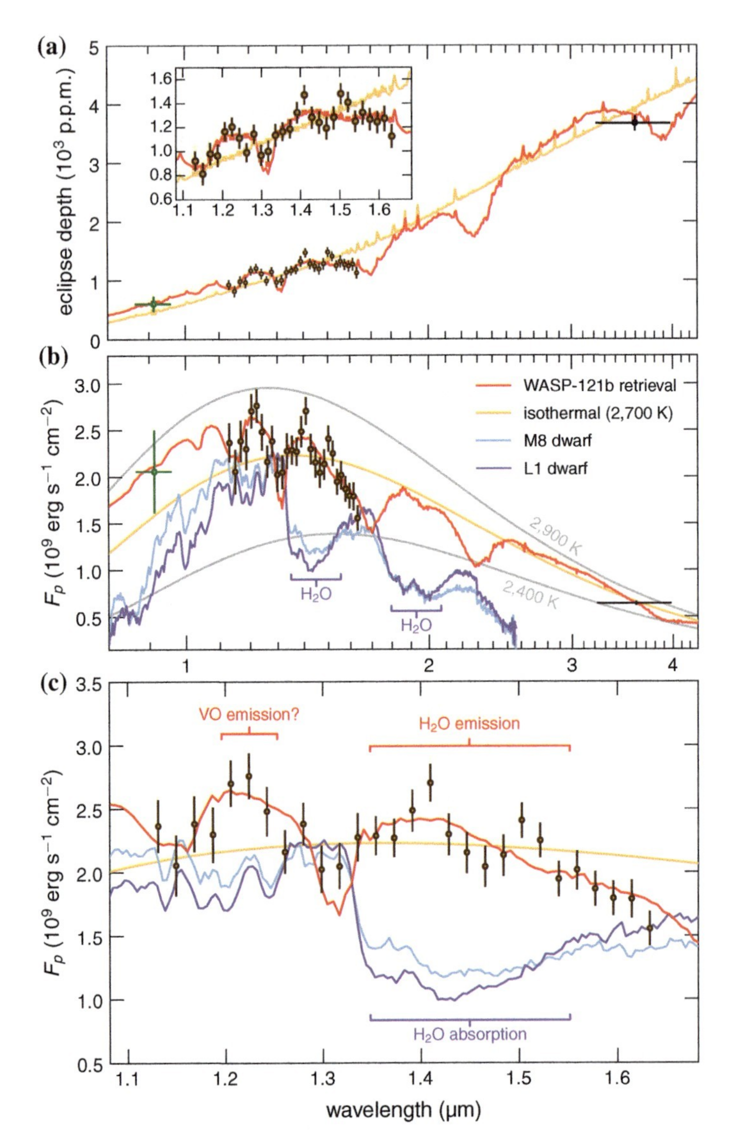

Fig. 1.9 Emission spectrum of WASP-121b from Evans et al. (2017). Shown is the measured HST, Spitzer, and ground-based eclipse depths and the 1-σ uncertainties, the horizontal error indicate the photometric bandpasses. The yellow lines show the best-fit isothermal blackbody spectrum with a temperature of 2,700 K, while the red lines show the best-fit atmospheric model from a retrieval analysis. The middle and lower panels show isolated planetary flux with the stellar contribution removed, and observed spectra for a M8 and L1 dwarf are shown for comparison (blue and purple lines), exhibiting H_2O absorption bands. For WASP-121b the H_2O band appears in emission. Spectra for 2,400 and 2,900 K blackbodies (grey lines) indicate the approximate temperature range probed by the data

near-IR, covering or muting the water absorption features (see Fig. 1.10). Most of the exoplanets characterized thus-far show some levels of clouds (Sing et al. 2016), though the strong diversity of cloud and haze covers found thus far indicates there will be a sizeable population hot Jupiters with largely clear atmospheres, especially in the infrared where the scattering opacity of hazes and clouds is likely to become greatly reduced.

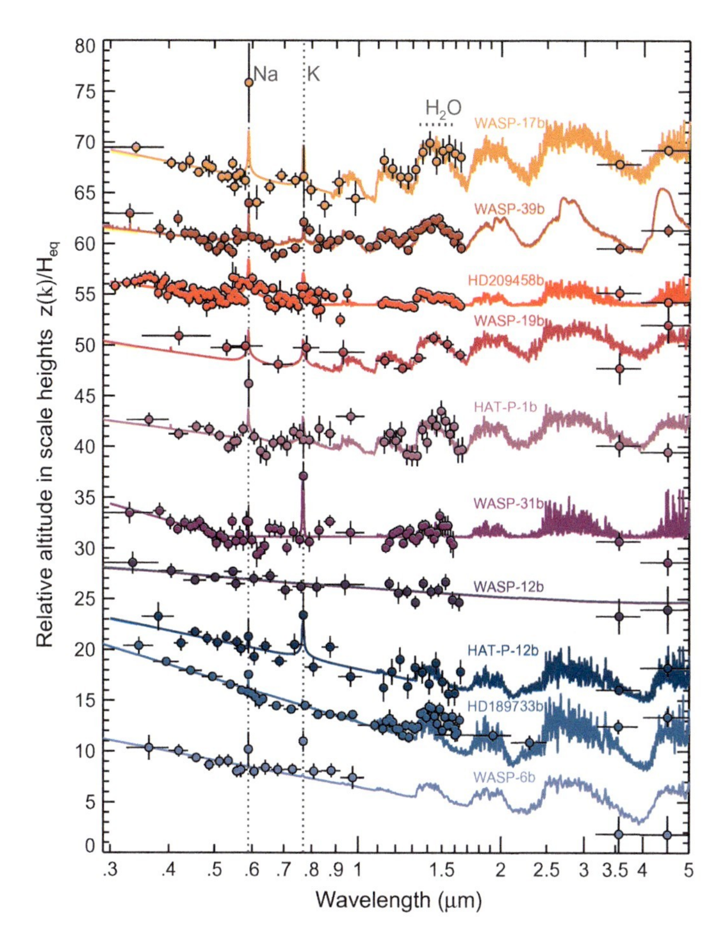

Fig. 1.10 Adapted from Sing et al. (2016). Shown is the HST/Spitzer transmission spectral sequence of hot-Jupiter survey targets, with data from Wakeford et al. (2018) for WASP-39b also included. Solid coloured lines show fitted atmospheric models and the prominent spectral features including Na, K and H_2O are indicated. The horizontal and vertical error bars indicate the transmission spectra wavelength bin and 1σ uncertainties, respectively. Planets with predominantly clear atmospheres (towards top) show prominent alkali and H_2O absorption, with infrared radii values commensurate or higher than the optical altitudes. Heavily hazy and cloudy planets (towards bottom) have strong optical scattering slopes, narrow alkali lines and H_2O absorption that is partially or completely obscured

1.3 Analysing Transmission Spectral Data

In the following sections, some of the basic data analysis procedures that have been developed to handle transit and eclipse spectrophotometry at high precision are reviewed. Included is a broad overview of the reduction steps, time series fitting methods, statistical tools, and methods to handle different noise sources. When pushing transiting exoplanet spectroscopy to very high (few ppm) levels, all potential sources of noise tend to matter, and even what may appear to be very minor effects can become important limiting factors.

1.3.1 Pre-observation Steps

An important but often overlooked aspect of observational astronomy and transiting exoplanet characterization is the steps one has to make and plan for well in advance of working with any new dataset. Most all good science starts with an idea, and a science question to investigate. For transiting exoplanet science, even fifteen years after the first atmospheric detection there are still no dedicated instruments designed from the beginning to perform the difficult task of obtaining the 10's or even 100's of ppm level photometric precisions across hour-long timescales necessary. This is in stark contrast to a dedicated instrument such as HARPS, which has proven $\sim$m/s radial velocity accuracies can be reliably achieved. As such, the planing and executing of transit/eclipse data still requires special care. Space-based data remains the gold standard, as the data quality is much more homogenous and the levels of precision demonstrated are much more reliably obtained. Nevertheless, a common mistake often made is to be overly optimistic or unrealistic in the levels of precision that can be achieved, and the size of potential atmospheric signatures. The current history of transiting exoplanet atmospheres has shown that more often than not, the atmospheric signal sizes observed are smaller than predicted, and the noise levels achieved are usually larger than photon limited (which exposure time calculators assume). Such was the case for Na on HD 209458b (Charbonneau et al. 2002) which was about $3\times$ smaller than predicted, and most of the H_2O features seen by HST to date as well have also been considerably smaller than predicted (e.g. Deming et al. 2013; Wakeford et al. 2013; Kreidberg et al. 2014b). Such has the case been with secondary eclipse measurements as well, with the Grillmair et al. (2008) Spitzer spectrum of HD 189733b showing H_2O features which were much smaller than earlier models (e.g. Burrows et al. 2008). Much of the smaller features are likely due to natural explanations such as clouds and hazes, which are still not robustly handled by most theoretical forward models given the complexity. For an observer, these aspects need to be kept in mind when planing one's observations to ensure detections can be made even if photon noise is not achieved and the signals sought for are smaller than expected. After all, the end goal is (or should be) to obtain robust, impactful results published in reputable journals, not just to collect data.

Basic Pre-Data Procedure Steps:

1. Have great idea to solve important science question, linking it to potential observations.
2. Check to see if the appropriate observations can be acquired, and with sufficient S/N. Do not be overly optimistic or unrealistic in what can be achieved, if so, even if telescope time is awarded and good data is acquired, luck will be needed to obtain impactful results.
3. Write a great telescope proposal(s).
4. Submit the proposal & convince a skeptical allocation committee.
   If rejected, revise proposal and resubmit next call if the case remains strong.
5. Plan the observations very carefully.
   a. Being paranoid of mistakes is often a good thing.
   b. Execute the observations very carefully.
   c. Do not underestimate the value of calibration frames.
   d. Make use of your collaborator's expertise.
   e. Maintain high photometric precisions over continuous multi-hour timescales. Most telescope operators are not used to the methods to obtain high photometric precisions (e.g. no dithering due to the relative nature of the measurement) and care must be taken to ensure the observations are executed properly.
6. Download/collect images.

1.3.2 Initial Calibration Overview

After obtaining a time-series dataset, the first data-reduction steps are much the same as any other method. However, before one begins reducing data it is a very good idea to inspect each of the data-frames in detail. A time-series dataset may contain hundreds or perhaps thousands of images. With these large numbers, it can be easy to overlook subtle issues which may not be obvious after aperture photometry is performed or the spectral trace is extracted. Stretching an image and adjusting the image contrast to view the high and low count level features can reveal possible issues such as bad pixels, and making time-series movies of the data or blinking frames can be a good method to reveal and get a feel for potential issues such as positional drifts, cosmic rays, and detector ghosts. Unexpected features may appear as well, as very unlikely asteroid or satellite crossing events have been noticed in HST time series data.

For HST WFC3 data, time-series spectral pixel maps can be utilised where a 2D histogram of the count levels at each pixel in a spectra are plotted vs time (e.g. Fig. 1.11). As positional drifts are a major cause of systematic errors in WFC3 spatially scanned data, these maps help reveal the extent of such drifts, and the presence of bad columns.

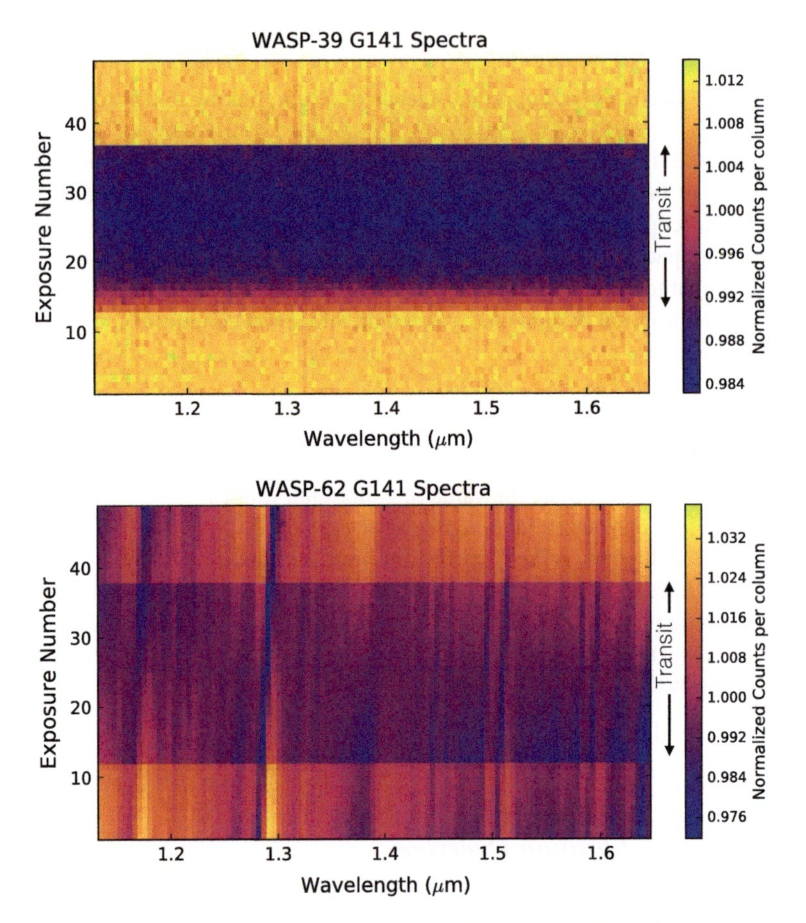

Fig. 1.11 Time-series spectral pixel maps for HST WFC3 G41 transit data of WASP-39b (top) and WASP-62b (bottom) used to visualise the overall data quality. Plotted is the normalized count level at each pixel in the spectra vs time during the transit observation. The transit event in both plots is evident by the drop in flux centred near the middle of both plots. The WASP-62b transit suffered from a guidestar problem which negatively impacted the telescope guiding, and large drifts on the order of several pixels can be seen. This problem negatively impacts the photometric quality of the spectroscopic channels, as is apparent by the systematic trends which drift in wavelength during the observation. An example of a good quality dataset is shown for WASP-39b (top) for comparison, which does not show such wavelength-dependent trends. Figures courtesy of H. Wakeford

The first reduction steps with modern 2D image or spectral data are essentially the same as all other traditional areas of astronomy. For CCD data, the images are typically:

- Trimmed of overscan regions, leaving the areas of the chip that contain useful data.
- Individual bias frames are combined.
- The flat-fields and science frames are processed to remove the overscan and average bias.

- Bad pixel maps are constructed.
- Flat-field images are combined and normalized.
  For time-series data, there is an important difference between low-frequency features (large scale trends) in a flat-field, and high-frequency pixel-to-pixel trends. A widely adopted method for most all time-series transit/eclipse data is to keep the point-spread-function of the telescope on the same pixel (or sub-pixel if possible) during the entire course of the observations. By doing so, low-frequency flat-field features are not important and largely do not impact the photometry. A transit or eclipse light curve is a differential measurement, and the absolute gain or count levels on the detector are not utilised as each light curve is individually normalized to the out-of-transit levels. Thus, many observers choose not to apply a flat-field correction, as it has been sometimes been seen to introduce noise (e.g. Gibson et al. 2017). A successful flat-field correction, however, can correct for high-frequency pixel-to-pixel variations. These smaller scale gain differences between neighbouring pixels can become important if the telescope pointing is not entirely stable during the night, or if there are significant seeing changes. In these cases a flat-field correction may prove important. Given the extremely high count levels of the science frames that are obtained during a transit, very large numbers of well exposed flat-field images are needed such that the photon noise levels per-pixel in the combined flat-field is comparable to the integrated photon noise levels per pixel in the science frames. Thus, be sure to obtain as many flat-field and bias images as possible, as it is unlikely taking one or two flat-field images would prove useful when a time-series dataset is aiming for high photometric precisions measuring a transit depth over perhaps hundreds or thousands of images.
- The science exposures are flat-fielded, and bad-pixels corrected if desired.
- Cosmic rays are cleaned.
- The spectra center is determined, trace defined, and 1D spectra extracted.
  In a 2D spectral image, the center of the PSF at each pixel along the cross-dispersion direction is determined, called the trace. The background region is then defined and the counts in the cross-dispersion direction are then summed in an aperture of a given size along the trace to add up the total counts for each wavelength-pixel in the spectrum, typically subtracting the background. The optimal aperture size and optimal background region have to be explored to optimise the time-series photometry at later stages.
- A wavelength solution is determined.
  Typically, arc lamp frames are gathered and the wavelength of specific emission lines are identified, which then provides a direct wavelength-to-pixel mapping. In the case of transit spectroscopy, stellar absorption features can also provide a direct and unambiguous identification of the wavelength (for example the Na D lines in optical spectra).

With space-based transit spectroscopy, one can proceed more or less directly from spectral extraction to fitting light curves. In the case of ground-based multi-object spectroscopy (MOS), the spectra of two or more stars must be extracted as a reference star is needed to correct each spectroscopic channel for the effects of Earth's atmosphere. The method of MOS is essentially an extension of differential photometry, and was initially applied by Bean et al. (2010) to observe the transmission spectra of GJ1214b with VLT FORS2. In MOS observations, the spectra of two or more stars are obtained using typically a mask or longslit, with the slit sizes specifically chosen to be very wide such that slit light losses are minimized or eliminated all together (see Fig. 1.12). Sizes of 10"+ or more are often used, with an example of Nikolov et al. (2016) using 22"× 90" sizes slits on VLT FORS2. The target and reference spectra must then be accurately wavelength calibrated, such that light curves at the same wavelengths can be binned and constructed for use in differential photometry (see Fig. 1.13). In principle, the reference star should have all the adverse effects from Earth's atmosphere also encoded in the light curves including seeing variations, transparency variations, and changing atmospheric extinction. Dividing the target star light curve by the reference star then largely subtracts out these features. In practice, other effects must also be dealt with such as instrument flexure and pointing drifts.

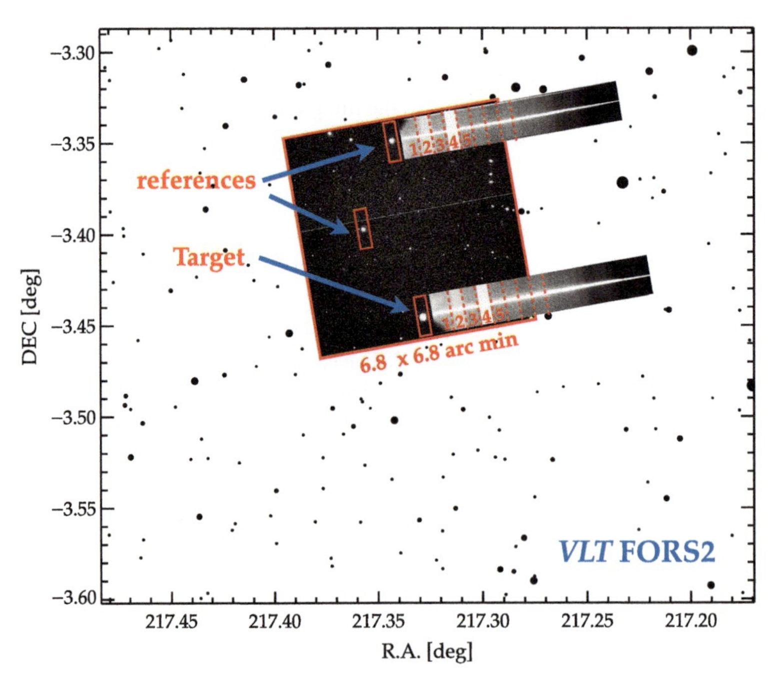

Fig. 1.12 Illustration of ground-based multi-object differential spectroscopy. Shown is the field of view of VLT FORS2 instrument along with the surrounding field of WASP-6b, overlaid with an acquisition image and a spectroscopic image. The red rectangles on the acquisition image show the large slits as they are projected on the sky, which encompass the target and a number of nearby reference stars. The numbered rectangles depict example wavelength bins for the target and reference star. Figure courtesy of N. Nikolov

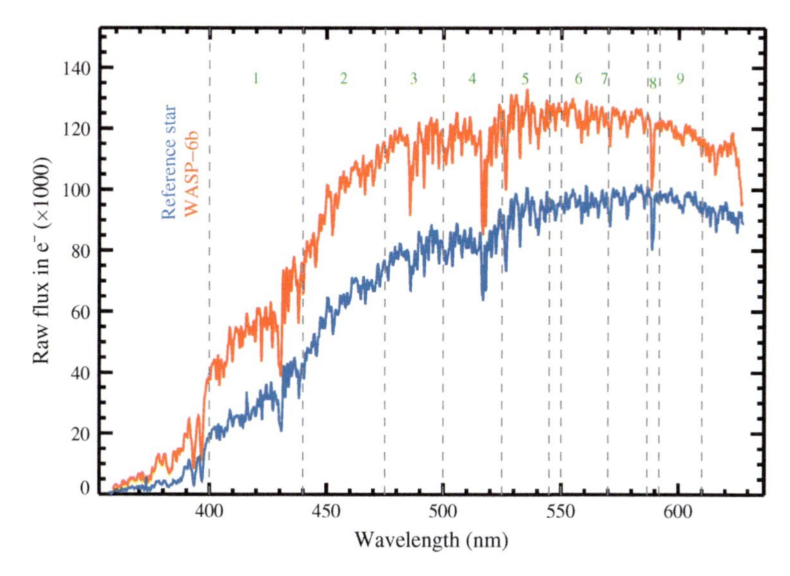

Fig. 1.13 Stellar spectra of WASP-6A and a target star, along with the bins used to create the spectroscopic channels transit light curves. Figure courtesy of N. Nikolov

1.3.3 Pre-light Curve Fitting

For most areas of astronomical observations, the data reduction steps would largely stop here and the scientific analysis would begin, though with transmission or emission spectroscopy of exoplanets, the job is just partially complete. The very important steps of fitting the transit light curves (along with systematic trends) for each spectroscopic channel still need to be performed. In most cases, the light curve fitting stage takes the most time and effort and it remains a non-trivial process which must be handled with care.

There are several important initial steps which need to be taken before fitting transit light curves:

- Calculate accurate time stamps for the time series images.
  Very high quality light curves can often have quantities like the central transit time measured to precisions of seconds or better, and other studies will often use the measured quantities to look for transit time variations or followup an exoplanet at a later date, making accurate time stamps very important. Indeed, care must be taken when combining different datasets from different observatories, as it is often hard to know exactly what clock standard was used. The commonly used Julian Date (JD) can be specified in several different time standards which are often unreported, complicating timing studies. Fortunately, Eastman et al. (2010) provides a useful tool to calculate the Barycentric Julian Date in the Barycentric Dynamical Time (BJD_{TDB}) standard, which has become a common practice in the exoplanet literature (http://astroutils.astronomy.ohio-state.edu).

With an accurate time-stamp, the projected separation between the planet and star, s, as a function of the orbital phase, ϕ, can be calculated as,

$$s(\phi) = \frac{a}{R_{star}} \sqrt{[\sin(2\pi\phi)]^2 + [\cos(i)\cos(2\pi\phi)]^2} \tag{1.8}$$

where a is the semi-major axis and i is the inclination of the orbit. $s(\phi)$ is a quantity provided as input to widely used transit models such as Mandel and Agol (2002).

- Align the spectra onto a common wavelength/pixel grid.
  Spectra are often seen to exhibit non-trivial shifts on the detector during the time series sequence. These shifts need to be accounted for if the spectra are to be placed on a common wavelength grid such that spectral bins can accurately be extracted. Without applying such a correction, fixed pixel bins of the spectral time series could contain a mix of wavelengths as the light from neighbouring pixels would contaminate the bin and degrade the light curves. Shifts between spectra can be easily measured using cross correlation procedures, and the spectra can be interpolated onto a common scale.
- Calculate limb darkening coefficient for the wavelength bin of interest.
  Limb darkening is an essential component determining the shape of a transit light curve, enhancing the U-shape of transit light curves due to the non-uniform flux profile across stellar discs. Thus, an accurate treatment of stellar limb darkening is critical when deriving precise planetary radii and measuring transmission spectra. The effects of limb darkening are typically parameterized using a specified functional form (or law), and theoretical limb-darkening coefficients (LDCs) are calculated using stellar models (e.g. Claret 2000). The most commonly used in exoplanetary transit work are:
  the linear law

$$\frac{I(\mu)}{I(1)} = 1 - u(1-\mu), \tag{1.9}$$

the quadratic law

$$\frac{I(\mu)}{I(1)} = 1 - a(1-\mu) - b(1-\mu)^2, \tag{1.10}$$

the three-parameter non-linear law,

$$\frac{I(\mu)}{I(1)} = 1 - c_2(1-\mu) - c_3(1-\mu^{3/2}) - c_4(1-\mu^2), \tag{1.11}$$

and the four-parameter non-linear law

$$\frac{I(\mu)}{I(1)} = 1 - c_1(1-\mu^{1/2}) - c_2(1-\mu) - c_3(1-\mu^{3/2}) - c_4(1-\mu^2), \tag{1.12}$$

where $I(1)$ is the intensity at the centre of the stellar disk, $\mu = cos(\theta)$ (where θ is the angle between the line of sight and the emergent intensity), while u, a, b, and

c_n are the LDCs. These laws can all be used along with the analytical transit light models of Mandel and Agol (2002) or Kreidberg (2015).
Fitting for LDC from the transit light curves is widely used, with the quadratic law most often adopted. However, there are degeneracies in fitting for the coefficients and without a proper treatment unphysical stellar intensities can result which can bias the results. This can happen especially for grazing transits as the full stellar disk (and its intensity profile) is not sampled during the transit. In addition, simple limb darkening laws can also do a poor job of reproducing a real stellar intensity profile. For many transmission spectral applications, theoretical stellar models have proven adequate for many transit light curve fits (e.g. Sing et al. 2011) and the latest 3D models (see Fig. 1.14 and Hayek et al. 2012; Magic et al. 2015) have improved upon many of the deficiencies seen in earlier 1D models (Knutson et al. 2007a; Sing et al. 2008). For transit spectroscopy, it is often recommended to fix the limb-darkening coefficients to their theoretical values, and inspect the fitted residuals to see how well the stellar models are performing, and fit for the coefficients if necessary.

1.3.4 Light Curve Fitting

When fitting for a spectroscopic transit or eclipse dataset, the first fits one typically performs is on the wavelength-integrated flux of the spectrum, which is called the white-light curve. The white-light curve fit helps provide the overall system parameters such as i, a/R_{star} and center of transit time T_0 as well as the average transit depth across the wavelength range of the spectrum.

In most all transit light curves to date, a model of the systematic trends (any non-transit/eclipse related phenomena which affects the light curve) must also be taken into account, whether they are of an instrumental or astrophysical nature. For example, in Spitzer IRAC transit photometry, it has been widely established the intra-pixel sensitivities and pointing jitter cause variations in the photometric light curves which must be modelled and removed (Morales-Calderón et al. 2006; Knutson et al. 2008). For HST STIS data, thermal breathing trends cause the point-spread-function (PSF) to change repeatedly for each 90 min spacecraft orbit around the Earth, producing corresponding photometric changes in the light curve which results in photometric changes in the light curve (Brown et al. 2001). Systematic errors are often removed by a parameterized deterministic model, where the non-transit photometric trends are found to correlate with a number n of external parameters (or optical state parameters, $\mathbf{x}$). These parameters describe changes in the instrument or other external factors as a function of time during the observations, and are fit with a coefficient for each optical state parameter, p_n, to model and remove (or detrend) the photometric light curves.

When including systematic trends, the total parameterized model of the flux measurements over time, $f(t)$, can be modelled as a combination of the theoretical transit

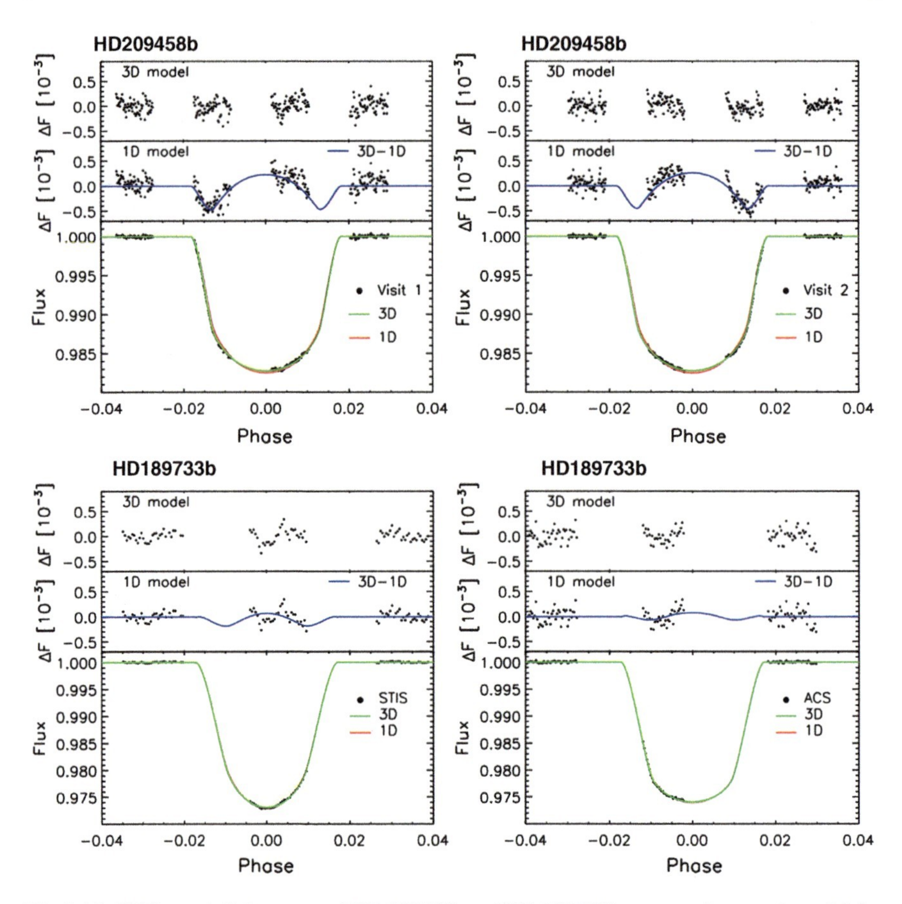

Fig. 1.14 HST transit light curves of HD 209458b and HD 189733b compared to transit model fits using 1D and 3D models (adapted from Hayek et al. 2012). (Top) For HD 209458b 1D models are unable to fully reproduce the transit shape, leading to a characteristic "w" shaped residual as seen in the middle panels for each fit and 3D stellar models do a better job of fitting the transit. (Bottom) The 1D models perform better for HD 189733b

model, $T(t, \theta)$ (which depends upon the transit parameters θ), the total baseline flux detected from the star, F_0, and the systematics error model $S(\mathbf{x})$ giving,

$$f(t) = T(t, \theta) \times F_0 \times S(\mathbf{x}). \tag{1.13}$$

In the case of HST STIS data, external detrending parameters including the 96 min HST orbital phase, ϕ_{HST}, the X_{psf} and Y_{psf} detector position of the PSF, and the wavelength shift S_λ of the spectra have been identified as optical state parameters (Sing et al. 2011). The optical state parameters must be properly normalized such that they do not contribute in changing the overall average system flux, and in the case of STIS data a fourth-order polynomial with ϕ_{HST} has been shown to sufficiently correct the instrument systematics such that $f(t)$ can be written as,

$$f(t) = T(t,\theta) \times F_0 \times (p_1\phi_{HST} + p_2\phi_{HST}^2 + p_3\phi_{HST}^3 + p_4\phi_{HST}^4 + 1) \times (p_5 S_\lambda + 1) \times (p_6 X_{psf} + 1) \times (p_7 Y_{psf} + 1). \quad (1.14)$$

Note that the systematics model can also be additive rather than multiplicative as in Eq. 1.14. Determining what systematic model to use, identifying parameters which successfully detrend light curves, and finding suitable functions while avoiding over-fitting remains a large non-trivial challenge when analysing transit light curves. This is especially true if the data collected does not have a significant history and standard practices determined when dealing with high precision photometric time-series measurements. Non-parametric methods have also been developed to model systematic models, and methods such as Gaussian processes (GP) have the benefit of not imposing a specific functional dependence on the optical state vectors (see Gibson et al. 2012a), and have become widely used.

When fitting transit models to the data, typically one begins by finding a minimum χ^2 solution (see Fig. 1.15) from routines such as the Levenberg-Marquardt (L-M) least squares method (Markwardt 2009), while Markov chain Monte Carlo techniques (MCMC) methods are useful for deriving robust error estimates that can account for complicated degeneracies between model parameters (Eastman et al. 2013; Foreman-Mackey et al. 2013).

The size of the photometric error bars are a critical, and often overlooked aspect of light curve fitting. Typically, the minimum χ^2 values and best-fit parameters are insensitive to the exact size of the errors (within reason). However, the inferred uncertainties on the best-fit model parameters themselves are dependent upon the size of the photometric errors/uncertainties given. In high-precision time-series photometry, it is often the case that the formal error bars are dominated by photon noise. However, for many datasets photon noise is rarely achieved and can even be factors of a few away from achieving those levels of precision. A common practice is to adopt photon-noise error bars when performing model selection, such that a given model $f(t)$ performance can be compared to the theoretical limit of the data, and a comparison can be made between the performance of different models. However, when a satisfactory model is found, the photometric error bars themselves are rescaled by the standard deviation of the residuals, and the model is re-fit. Alternatively, for methods such as MCMC that optimise the log-likelihood, the photometric/white noise level can be fit as a free parameter itself. Rescaling the error bars in these ways helps take into account noise sources that are difficult to account for in standard pipeline reduction routines, and the final uncertainties will be underestimated if this step (or a similar procedure) is not taken, unless of course photon noise precisions were achieved. In addition, the presence of time-correlated red noise must also be considered, with the binning technique (Pont et al. 2006) and wavelets (Carter and Winn 2009) two recommended methods. With the photometric error bars set to realistic values, and red noise properly handled, MCMC routines (e.g. Eastman et al.

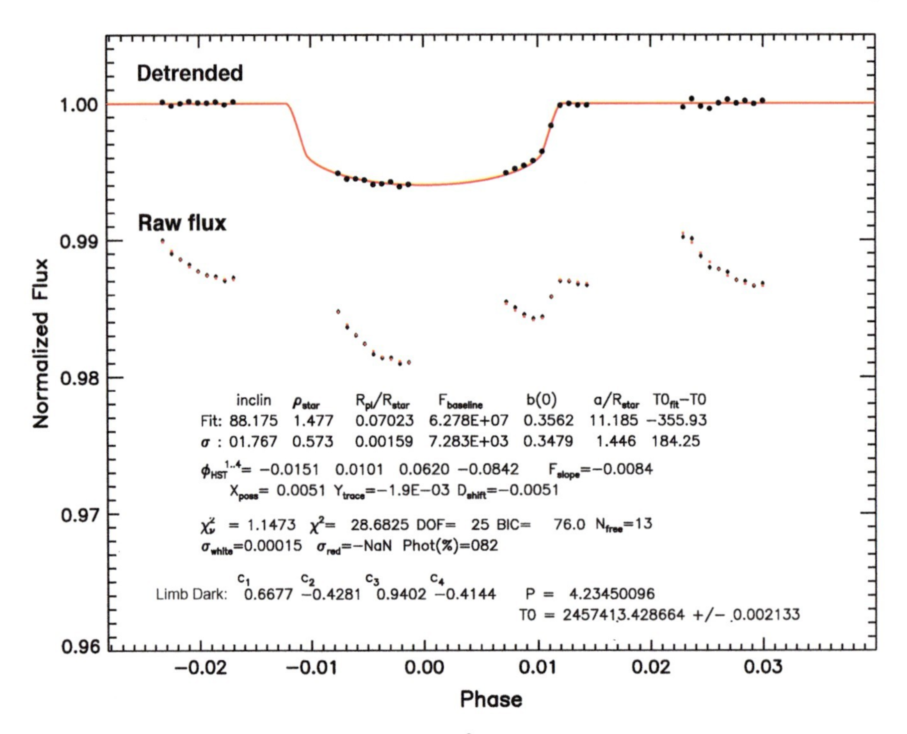

Fig. 1.15 Example transit light curve minimum χ^2 fit to HST STIS data of HAT-P-26b (Wakeford et al. 2017). Top shows the detrended normalized light data (black points) with the transit light curve model (red line), middle shows the raw flux which includes the instrument systematic trends along with the best-fit model (red points), an arbitrary offset has been applied for clarity. The best-fit model parameters and several statistics of the model are also indicated. In this example, the fit achieves precisions of 82% the theoretical photon noise levels

2013; Foreman-Mackey et al. 2013.) can then be run to provide the full posterior distribution and marginalised parameter uncertainties (see Fig. 1.16).

After the white light curve is fit, the spectroscopic channels are then determined and transmission or emission spectra constructed. During this stage, it is typically only the transit or eclipse depth that is allowed to vary along with the systematics model in the fits. All other parameters tend to be fixed to those values adopted for the white light curve, as they are not expected to vary across spectroscopic channels. This includes the semi major axis a, orbital inclination i and transit/eclipse mid-time. An important exception is the limb-darkening coefficients, which as noted above must be carefully determined for each spectroscopic channel individually as they are wavelength-dependent.

Below are a few tips to consider regarding transit observations:

1. Observe whole time series on the same pixel, sub-pixel if possible.
2. Choose a detector setting (e.g. gain) to get as many counts as reasonably possible per image in order to improve the relative photometric precision.

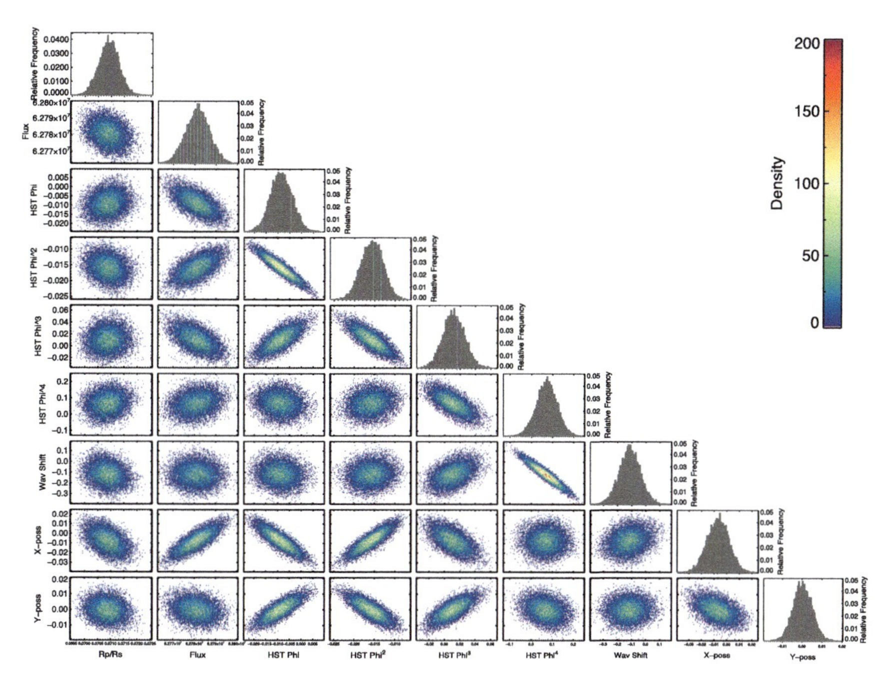

Fig. 1.16 Example MCMC posterior distribution using of a transit light curve model fit to HST STIS HAT-P-26b data (Wakeford et al. 2017)

3. Readnoise is almost never important, as we tend to observe bright targets, so the error budget is dominated by photon noise.
4. Red noise is always important to consider.
5. For MOS, wide slits/apertures are needed to minimise differential slit losses between the target and reference stars.
6. Have a uniform detector.
7. Be mindful of detector electronics which can introduce noise.
8. Get a long enough baseline before and after the transit/eclipse to properly measure the depth.
9. As much as possible, have the out-of-transit/eclipse baseline sample similar systematics to those in-transit/eclipse.
10. Ingress and egress are not very useful to measure the transit/eclipse depth but are needed to measure a/R_{star}, T_0, and the inclination.
11. Avoid or correct for non-linearities in the detector.
12. Reference stars need to be as similar as possible to the target in type & magnitude (preferably within 1 mag, and spectral sub-type).
13. Near-IR is hard from the ground, due to terrestrial opacity sources such as water, but not impossible.
14. Philosophy differs in flat-fielding (I suggest gathering 100's, test to see if they improve things or not).

15. Cosmic Rays are important.
16. Visualise your time series data (e.g. with a movie) to see all the ways your spectra/photometry changes in time.
17. Be wary of hidden companions that can dilute the flux of the brighter star you're interested in.
18. Be wary of unphysical results.

1.4 Interpreting a Transmission Spectrum

In the following section, an analytic transmission spectrum formula is derived from first principles. Details of this derivation can be found across Fortney et al. (2005), Lecavelier des Etangs et al. (2008), de Wit and Seager (2013), Bétrémieux and Swain (2017) and Heng and Kitzmann (2017). Given that a transmission spectrum is not a typical astrophysical measurement, the derivation helps to illustrate how the spectrum depends upon several key parameters (such as the temperature and molecular abundances), what quantities can be derived from a transmission spectra, and the nature of modelling degeneracies.

1.4.1 Analytic Transmission Spectrum Derivation

The flux blocked during transit including the planet and its atmosphere is given by the ratio of the areas,

$$\frac{\Delta f}{f} = \frac{\pi R_{pl}^2 + A}{\pi R_{star}^2}, \tag{1.15}$$

where A is the effective area of the annular region of the atmosphere observed during transit. The contribution of the atmosphere, A, is calculated by integrating the absorptivity of the atmosphere from a reference planetary radius, R_{pl}, up to the top of the atmosphere along the radial coordinate direction r' and is given by,

$$A = \int_{R_{pl}}^{\infty} (1 - T) 2\pi r' dr' \tag{1.16}$$

where T is the transmittance which is the fraction of radiation that is transmitted through a given layer of atmosphere. The transmittance is related to the optical depth τ using the Beer-Lambert law,

$$T = e^{-\tau}, \tag{1.17}$$

and the optical depth in turn can also be written in terms of the cross section of absorbing species $\sigma_{abs}(\lambda)$, its number density n_{abs}, and integrated along the slant transit geometry (direction $\hat{x}$),

$$\tau = \int_{-\infty}^{+\infty} \sigma_{abs}(\lambda) n_{abs} dx, \tag{1.18}$$

$$T = e^{-\int_{-\infty}^{+\infty} \sigma_{abs}(\lambda) n_{abs} dx}. \tag{1.19}$$

As the number density of an atmosphere drops with altitude radially (direction $\hat{r'}$), a simplifying approximation can be made by assuming the atmosphere is isothermal with temperature T, an ideal gas with pressure $P = k_B \rho T/\mu$, where ρ is the density, and in hydrostatic balance,

$$dP = -\rho g dr. \tag{1.20}$$

Substituting ρ from the ideal gas law and integrating Eq. 1.20 then gives the Barometric formula which relates the pressure between two points (labeled here as 0 and 1) with the altitude difference, $P_1 = P_0 e^{-r/H}$, and pressure scale height H or equivalently the number density between two altitude points,

$$n_1 = n_o e^{-r/H}. \tag{1.21}$$

Typically, the absorbing species in a transmission spectra is a minor component of the gas, so we can relate the mixing ratio of the absorbing minor species, ξ_{abs}, to the total gas number density, by $n_{abs} = n_o \xi_{abs}$. Substituting Eq. 1.21 into Eq. 1.18 and evaluating the optical depth at the reference planetary radius and pressure (R_{pl}, P_0) gives,

$$\tau_0 = \xi_{abs} \sigma_{abs}(\lambda) \int_{-\infty}^{+\infty} \frac{P_0}{k_B T} e^{-r'/H} dx, \tag{1.22}$$

where we have made the assumption that the cross section $\sigma_{abs}(\lambda)$ does not depend upon the atmospheric pressure. The $\hat{x}$ and $\hat{r'}$ coordinates can be related by the Pythagorean theorem (see Fig. 1.17),

$$(R_{pl} + r)^2 = R_{pl}^2 + x^2 \tag{1.23}$$

$$x^2 = 2R_{pl} r + r^2, \tag{1.24}$$

and further simplified assuming planetary radius is much larger than the atmospheric altitude ($R_{pl} >> r$) which gives

$$r = \frac{x^2}{2R_{pl}}. \tag{1.25}$$

Substituting r into Eq. 1.22 gives,

$$\tau_0 = \xi_{abs} \sigma_{abs}(\lambda) \int_{-\infty}^{+\infty} \frac{P_0}{k_B T} e^{-x^2/2R_{pl}H} dx. \tag{1.26}$$

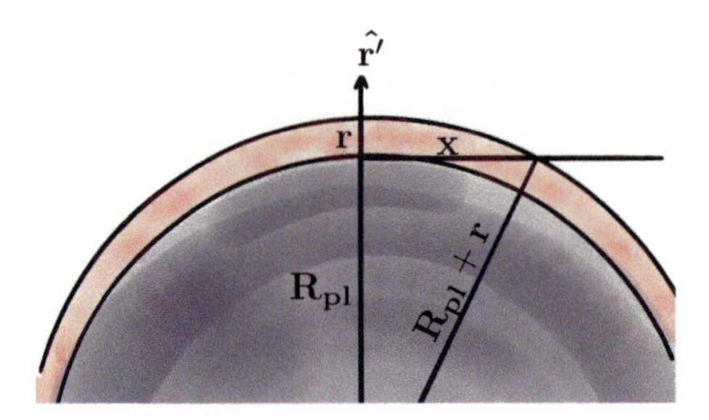

Fig. 1.17 Illustration of the transit geometry. The bulk radius of the planet, R_{pl} (grey) is illustrated along with an atmospheric layer (red) of thickness, r. Stellar light passes in slant transit geometry from the terminator through the atmospheric layer a distance x. The radial coordinate $\hat{r'}$ is also indicated with its origin at the planet center

The integral can be analytically evaluated as, $\int_{-\infty}^{+\infty} e^{-u^2} du = \sqrt{\pi}$, so the reference optical depth becomes,

$$\tau_0 = \xi_{abs}\sigma_{abs}(\lambda)\frac{P_0}{k_B T}\sqrt{2\pi R_{pl} H}. \tag{1.27}$$

From Eq. 1.27, it can be seen that the optical depth depends on the pressure (i.e. $\tau \propto P$), and again using the Barometric formula the optical depth at an arbitrary pressure P can be written as,

$$\tau = \xi_{abs}\sigma_{abs}(\lambda)\frac{P}{k_B T}\sqrt{2\pi R_{pl} H}. \tag{1.28}$$

Before we can evaluate Eq. 1.16, we need a relation between the radial coordinate $\hat{r'}$ and optical depths τ and $d\tau$. Using the Barometric formula in terms of pressure, and the fact that the optical depth is proportional to pressure gives,

$$\frac{\tau}{\tau_0} = \frac{P}{P_0} = e^{-r/H}, \tag{1.29}$$

$$\tau = \tau_0 e^{-r/H}, \tag{1.30}$$

the altitude difference, r, is then related to the reference radius of the planet and radial coordinate r' as $r = r' - R_{pl}$, thus

$$\tau = \tau_0 e^{-(r'-R_{pl})/H}, \tag{1.31}$$

$$\ln\left(\frac{\tau}{\tau_0}\right) = -\frac{r'}{H} + \frac{R_{pl}}{H}, \tag{1.32}$$

$$r' = R_{pl} - H \ln\frac{\tau}{\tau_0}, \tag{1.33}$$

and

$$dr' = -\frac{H}{\tau} d\tau. \tag{1.34}$$

Substituting these values r' of dr' into Eq. 1.16 along with Eqs. 1.28 and 1.17 then gives,

$$A = 2\pi \int_{\tau_0}^{0} (1 - e^{-\tau}) \left(R_{pl} - H \ln \frac{\tau}{\tau_0} \right) \left(-\frac{H}{\tau} d\tau \right), \tag{1.35}$$

where the integration limits have also been converted into the appropriate optical depth limits. After a slight rearrangement, Eq. 1.35 can be written

$$A = 2\pi H R_{pl} \int_{0}^{\tau_0} \left(\frac{1 - e^{-\tau}}{\tau} \right) \left(1 + \frac{H}{R_{pl}} \ln \frac{\tau_0}{\tau} \right) d\tau, \tag{1.36}$$

which can then be further simplified assuming $H/R_{pl} << 1$ so a term can be neglected to give,

$$A = 2\pi H R_{pl} \int_{0}^{\tau_0} \left(\frac{1 - e^{-\tau}}{\tau} \right) d\tau. \tag{1.37}$$

The integral can be found in Chandrasekhar (1960),

$$E_1 = -\gamma - \ln \tau_0 + \int_{0}^{\tau_0} \left(\frac{1 - e^{-\tau}}{\tau} \right) d\tau, \tag{1.38}$$

where γ is the Euler-Mascheroni constant (γ =0.577215664901...) and E_1 is an exponential integral which is a transcendental function. Thus, Eq. 1.37 becomes,

$$A = 2\pi H R_{pl} [E_1 + \gamma + \ln \tau_0]. \tag{1.39}$$

During a transit, we measure the transit-depth radius of the combined planet and atmosphere. Thus to relate Eq. 1.39 to the transit radius, we can set the measured transit depth altitude r_{eq} so it produces the same absorption depth as the planet with its translucent atmosphere occulting area, A, giving,

$$A = 2\pi R_{pl} r_{eq} = 2\pi H R_{pl} [E_1 + \gamma + \ln \tau_0], \tag{1.40}$$

$$r_{eq} = H[E_1 + \gamma + \ln \tau_0]. \tag{1.41}$$

As τ_0 is the optical depth an arbitrary reference pressure-altitude level, we can set this value to be very large ($\tau_0 >> 1$) to use the limit where $E_1 \rightarrow 0$ as $\tau_0 \rightarrow \infty$,

$$r_{eq} = H[\gamma + \ln \tau_0]. \tag{1.42}$$

Equation 1.42 can then be rearranged and the radius converted to optical depth (Eq. 1.34) to figure out the equivalent optical depth τ_{eq} where the transmission spectrum becomes optically thick in slant transit geometry and corresponds to the measured transit-depth radius giving,

$$r_{eq}/H = \gamma + \ln \tau_0, \tag{1.43}$$

$$e^{-r_{eq}/H} = \tau_{eq}/\tau_0 = e^{-\gamma - \ln \tau_0}, \tag{1.44}$$

$$\tau_{eq}/\tau_0 = e^{-\gamma}/\tau_0, \tag{1.45}$$

$$\tau_{eq} = e^{-\gamma} = 0.561459. \tag{1.46}$$

Lecavelier des Etangs et al. (2008) first derived τ_{eq} numerically, where $\tau_{eq} = 0.56$ is seen to be an accurate approximation (given the terms we have neglected) for most planetary atmospheres as long as (30> R_{pl}/H >300). Finally, we can substitute Eqs. 1.46 and 1.27 into Eq. 1.42 to give,

$$r_{eq} = H\left[-\ln \tau_{eq} + \ln\left(\xi_{abs}\sigma_{abs}(\lambda)\frac{P_0}{k_B T}\sqrt{2\pi R_{pl} H}\right)\right], \tag{1.47}$$

$$r_{eq} = H\left[\ln\left(\frac{\xi_{abs}\sigma_{abs}(\lambda)P_0}{\tau_{eq}}\sqrt{\frac{2\pi R_{pl} H}{k_B^2 T^2}}\right)\right], \tag{1.48}$$

and r_{eq} relabelled as $z(\lambda)$ which then derives the Lecavelier des Etangs et al. (2008) transmission spectrum formula,

$$z(\lambda) = H \ln\left(\frac{\xi_{abs}\sigma_{abs}(\lambda)P_0}{\tau_{eq}}\sqrt{\frac{2\pi R_{pl}}{k_B T \mu g}}\right). \tag{1.49}$$

1.4.2 Analytic Transmission Spectrum Applications

The first application of Eq. 1.49 was to interpret the transmission spectrum of HD 189733b from Pont et al. (2008) in which atmospheric haze on an exoplanet was first discovered. If the cross section $\sigma_{abs}(\lambda)$ is known, then the altitude difference between two wavelengths, dz, allows the pressure scale height H to be directly measured, which directly leads from Eq. 1.49 to,

$$T = \frac{\mu g}{k_B}\left(\frac{d \ln \sigma}{d\lambda}\right)^{-1}\frac{dz(\lambda)}{d\lambda}. \tag{1.50}$$

This temperature measurement will be accurate in the case where the absorbing species in the transmission spectra has been robustly identified, such as is often the case for Na, K, or H_2O, and when the mean molecular weight of the atmosphere is also known. Thus, the terminator temperatures in hot Jupiters can often be accurately measured, given the atmosphere is H/He dominated; though that may not be the case for super-Earths which could have much heavier non-H/He secondary atmospheres.

For HD 189733b, the slope of the transmission spectra indicated a scattering slope, and the cross section could be assumed to follow a power law of index α, such that $\sigma = \sigma_0(\lambda/\lambda_0)^\alpha$. In this case,

$$\alpha T = \frac{\mu g}{k_B} \frac{dR_{pl}(\lambda)}{d \ln \lambda}, \tag{1.51}$$

making the transmission spectrum slope proportional to the product αT. In the case of pure Rayleigh scattering, $\alpha = -4$ and the temperature can be derived, though in general if the pure Rayleigh scattering is not apparent, then the constrained quantity will be αT and a degeneracy will exist between the power law index and the atmospheric temperature.

Equation 1.49 can be used to straightforwardly make an entire optical transmission spectrum for a hot Jupiter, given that theoretical models have shown a typical hot Jupiter (~1200 K) will be dominated by Na, K, and Rayleigh scattering (Seager and Sasselov 2000; Hubbard et al. 2001; Brown 2001). Na and K are both doublets, so by including only four absorption lines and a scattering component, the majority of an hot Jupiter optical transmission spectrum can be modelled.

For largely clear atmospheres, Na and K can both exhibit large pressure-broadened wings which will dominate the optical opacity. These wings can be calculated analytically using a Voigt line profile, $H(a, u)$, and statistical theory, which predicts the collision-broadened alkali line shapes will vary with frequency ν as $(\nu - \nu_0)^{-3/2}$ outside of an impact region, which lies between the line centre frequency ν_0 and a detuning frequency $\Delta\sigma$ away from ν_0.

Within the impact region, the cross-sections for both the sodium and potassium doublets can then calculated following Burrows et al. (2000) and Iro et al. (2005) as,

$$\sigma(\lambda) = \frac{\pi e^2}{m_e c} \frac{f}{\Delta\nu_D \sqrt{\pi}} H(a, u), \tag{1.52}$$

where f is the absorption oscillator strength of the spectral line, m_e is the mass of the electron, e the electron charge. The Voigt profile $H(a, u)$ is defined in terms of the Voigt damping parameter a and a frequency offset u. The frequency offset is calculated as $u = (\nu - \nu_0)/\Delta\nu_D$, where $\Delta\nu_D$ is the Doppler width given by $\Delta\nu_\mathrm{D} = \nu_0/c\sqrt{2kT/\mu_\mathrm{Na,K}}$, with c the speed of light and $\mu_\mathrm{Na,K}$ the mean molecular weight of sodium or potassium. The damping parameter is given by $a = \Gamma/(4\pi\Delta\nu_\mathrm{D})$, where the transition rate Γ is calculated following

$$\Gamma = \gamma + \Gamma_\mathrm{col}, \tag{1.53}$$

where γ is the spontaneous decay rate and Γ_col is the half-width calculated from classical impact theory. Assuming a Van der Waals force gives,

$$\Gamma_\mathrm{col} = 0.071(T/2000)^{-0.7}\ \mathrm{cm^{-1}atm^{-1}} \tag{1.54}$$

for Na and

$$\Gamma_{\mathrm{col}} = 0.14(T/2000)^{-0.7}\ \mathrm{cm}^{-1}\mathrm{atm}^{-1} \tag{1.55}$$

for K (Burrows et al. 2000; Iro et al. 2005).

Outside of the impact region, the $(\nu - \nu_0)^{-3/2}$ power-law line shape is truncated using an exponential cutoff term of the form $e^{-qh(\nu-\nu_0)/kT}$, where h is Planck's constant and q a parameter of order unity, to prevent the line wing opacity from becoming overly large at large frequency separations. The detuning frequency, $\Delta\sigma$, can be estimated from Burrows et al. (2000) using

$$\Delta\sigma = 30(T/500\ \mathrm{K})^{0.6}\ \mathrm{cm}^{-1}. \tag{1.56}$$

for the sodium doublet and

$$\Delta\sigma = 20(T/500\ \mathrm{K})^{0.6}\ \mathrm{cm}^{-1}. \tag{1.57}$$

for the potassium doublet.

The wavelength-dependent total cross sections of the sodium or potassium D1 and D2 doublet can then be summed together, $\sigma(\lambda)_{Na,K} = \sigma(\lambda)_{D1} + \sigma(\lambda)_{D2}$, and the sodium and potassium opacities can also be summed together along with their abundances into Eq. 1.49 as,

$$\xi_{abs}\sigma_{abs} = \xi_{Na}\sigma_{Na} + \xi_K\sigma_K. \tag{1.58}$$

Finally, a Rayleigh scattering component can also be added to the alkali lines, given that $\sigma_0 = 2.52 \times 10^{-28}$ cm^2 at λ_0 =750 nm for molecular hydrogen and for a hot Jupiter $\xi_{H_2} \sim 1$.

A comparison between the analytic transmission spectrum (Eq. 1.49) and a numerical model from Fortney et al. (2010) is shown in Fig. 1.18. The analytic model reproduces the Na and K lines profiles well, and certainly better than the accuracy of any transmission spectral data to date. The small differences seen are mainly due to the inclusion of molecules such as H_2O in the Fortney et al. (2010) model, which can be seen as weak lines near the K doublet, and the inclusion of weaker Na/K lines (e.g. 0.4 μm) which have not been added in this example analytic model.

The analytic transmission spectrum (Eq. 1.49) can easily be used to fit data in a retrieval model exercise (e.g. Sing et al. 2015), which given the few parameters and analytic nature is very fast making it highly conducive to Markov chain Monte Carlo techniques (MCMC). In Fig. 1.19 the WASP-6b optical HST data from Nikolov et al. (2015) is well fit using only four parameters: the abundances of Na and K, as well as the temperature and baseline planetary radius.

In general, molecular species can also be modeled with the analytic transmission spectrum (Eq. 1.49) as well, in this case the number of spectral lines jumps from four up to 10^9 or 10^{10} depending on the line list and species in question, which dramatically increases the computational burden. Nevertheless, even if one uses a

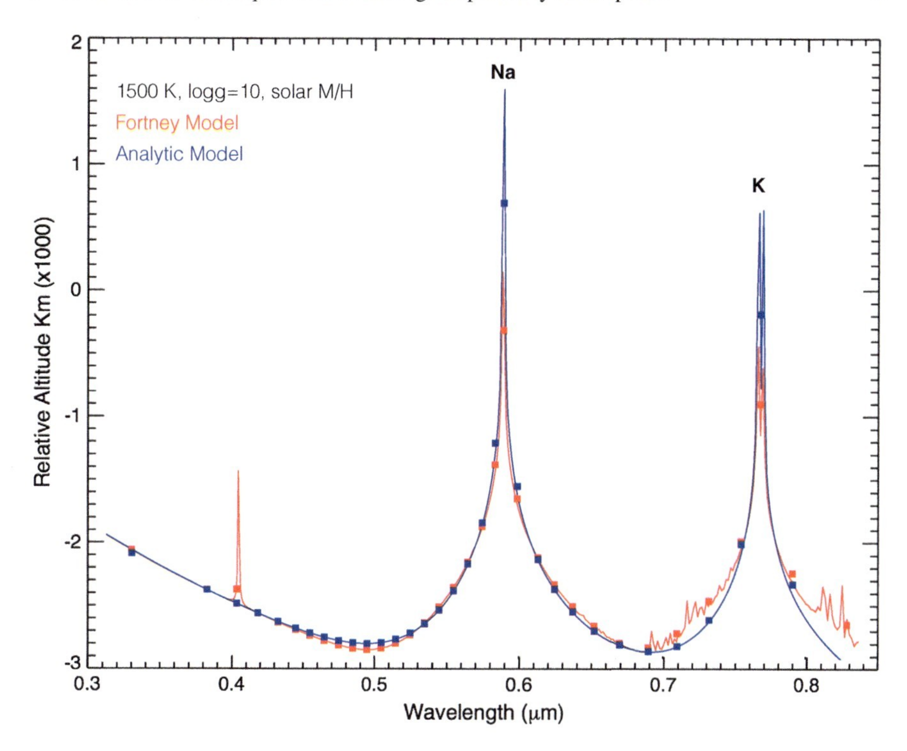

Fig. 1.18 Comparison between an analytic transmission spectrum model for a 1500 K hot Jupiter calculated using Eq. 1.49 (blue) and a model from Fortney et al. (2010)

fully numerical tool to calculate and model transmission spectra, it is a good idea to keep Eq. 1.49 in mind such that a physical intuition can be preserved.

A further consequence of Eq. 1.49 is that abundance ratios can be very precisely measured in a transmission spectrum. This follows from measuring the altitude difference in the spectra between the two (or more) wavelengths where the two species dominate. For example, taking the difference in altitudes between the transmission spectra at the Na core wavelengths, z_{Na}, and potassium core wavelengths z_K,

$$z_{Na} - z_K = H \ln\left(\frac{\xi_{Na}\sigma_{Na}(\lambda)P_0}{\tau_{eq}}\sqrt{\frac{2\pi R_{pl}}{k_B T \mu g}}\right) - H \ln\left(\frac{\xi_K \sigma_K(\lambda)P_0}{\tau_{eq}}\sqrt{\frac{2\pi R_{pl}}{k_B T \mu g}}\right), \tag{1.59}$$

which can be simplified to,

$$\frac{z_{Na} - z_K}{H} = \ln\left(\frac{\xi_{Na}\sigma_{Na}(\lambda)P_0}{\tau_{eq}}\right) - \ln\left(\frac{\xi_K \sigma_K(\lambda)P_0}{\tau_{eq}}\right), \tag{1.60}$$

$$\frac{z_{Na} - z_K}{H} = \ln\left(\xi_{Na}\sigma_{Na}(\lambda)\right) - \ln\left(\xi_K \sigma_K(\lambda)\right), \tag{1.61}$$

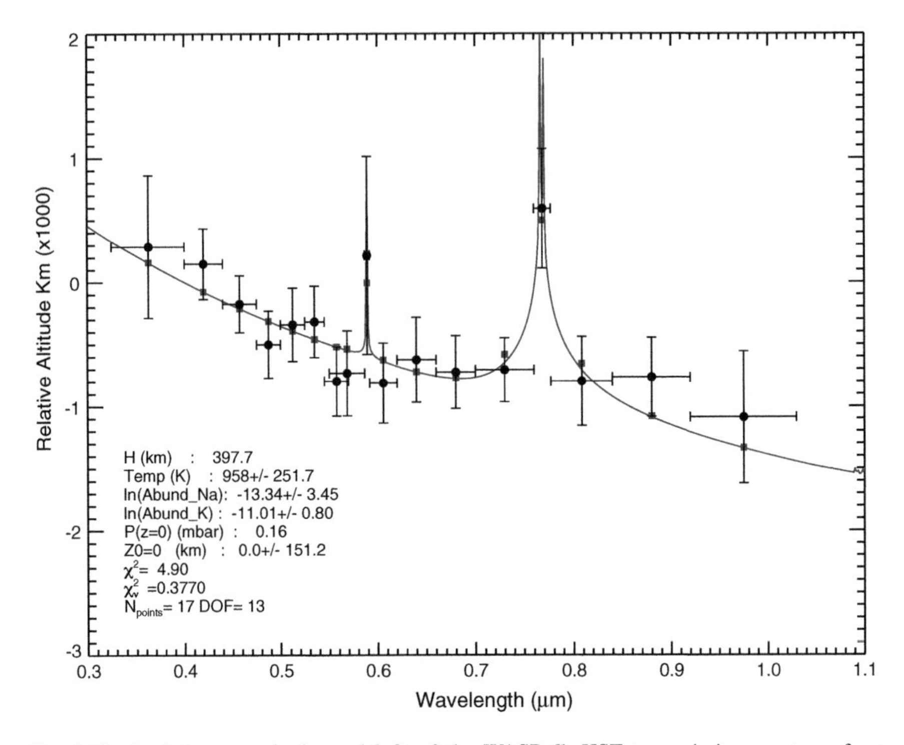

Fig. 1.19 Analytic atmospheric model fit of the WASP-6b HST transmission spectrum from Nikolov et al. (2015). Four parameters are fit, the temperature, Na and K abundances, and a baseline reference planet radius. The best-fit parameters and χ^2 statistics are indicated on the plot

$$e^{\frac{z_{Na}-z_K}{H}} = \frac{\xi_{Na}\sigma_{Na}(\lambda)}{\xi_K\sigma_K(\lambda)}, \tag{1.62}$$

$$\frac{\xi_{Na}}{\xi_K} = \frac{\sigma_K}{\sigma_{Na}} \exp\left(\frac{z_{Na}-z_K}{H}\right). \tag{1.63}$$

Almost all of the constants cancel, including importantly P_0 and R_{pl}, and the measured transit difference between the wavelengths of the two species directly translates into the abundance ratios between the two species. In the near-future, this aspect of transmission spectroscopy should lead to well measured C/O ratios in planetary atmospheres given the ratios of the dominant species (H_2O, CO_2, CO, & CH_4) may all be determined, e.g.

$$\frac{\xi_{H_2O}}{\xi_{CO}} = \frac{\sigma_{CO}}{\sigma_{H_2O}} \exp\left(\frac{z_{H_2O}-z_{CO}}{H}\right). \tag{1.64}$$

with early attempts of this measurement found in Désert et al. (2009) using Spitzer data.

1.4.3 Transmission Spectrum Degeneracies

An important degeneracy is apparent from Eq. 1.49 as the transmission spectrum is determined by the quantity of $\xi_{abs} P_0$. In general, the baseline pressure P_0 at the reference planetary radius is not known. Thus, it is typically difficult to measure absolute abundances in a transmission spectrum, which will be limited by the degeneracy between the abundance and baseline pressure. The degeneracy can also be re-cast as a degeneracy between ξ_{abs} and the reference planetary radius R_{pl}. Further discussion of degeneracies can be found in Benneke and Seager (2012) and Heng and Kitzmann (2017).

One way to lift the $\xi_{abs} - P_0$ degeneracy is to identify Rayleigh scattering by molecular H_2. Models such as those shown in Fig. 1.18 predict for clear hot Jupiter atmospheres that short-ward of about 0.5 μm, the H_2 molecular scattering should dominate the opacity. If this is the case, then the abundance of H_2 can be well approximated given $\xi_{H_2} \sim 1$, and the reference pressure P_0 at $z = 0$ can be determined. From Eq. 1.49, in the H_2-Rayleigh spectral region, the pressure P_0 at an altitude corresponding to the radius at wavelength λ_0 is

$$P_0 = \frac{\tau_{eq}}{\sigma_0} \sqrt{\frac{k_B T \mu g}{2\pi R_{pl}}}, \tag{1.65}$$

where σ_0 is the Rayleigh scattering cross section at λ_0. If P_0 can be determined from this method, the absolute abundances of all species identified in a transmission spectra (including molecular species at longer infrared wavelengths) can then be determined. In practice, hazes and clouds have often been observed to mask the H_2 Rayleigh scattering signatures. However, even if this is the case the short-wavelength region is still an important wavelength region to measure, as this data can constrain the cloud properties and rule out large classes of models, thereby constraining parameter space and limiting the $\xi_{abs} - P_0$ degeneracy.

In addition to the $\xi_{abs} - P_0$ degeneracy, the transmission spectra is also seen from Eq. 1.49 to scale (to first order) with the pressure scale height $H = k_B T/\mu g$. The H itself can typically be well determined from a well-measured transmission spectra. For hot Jupiter exoplanets which also have a measured mass, the surface gravity g is also known and μ can safely be assumed to be dominated by a H/He mixture leaving just the temperature as the unknown quantity that can be measured from the transmission spectrum. However, for non-gas giant exoplanets the molecular weight of the atmosphere can be unknown to perhaps an order of magnitude or more given atmospheres dominated for example by N_2, H_2O, or CO_2 are feasible. In addition, it is also more challenging to measure the mass of small exoplanets via the radial velocity method, which can also hinder a precise determination of g. Thus, in these cases there will be a large $T - g - \mu$ degeneracy, and the individual quantities will be difficult to constrain with the transmission spectrum alone.

Acknowledgements David K. Sing acknowledges the support from the meeting organisers for travel and accommodation at the meeting. DKS also gives thanks to Tom Evans, Nikolay Nikolov, Kevin Stevenson, and Hannah Wakeford for the use of several figures. DKS gives further thanks to Aarynn Carter, Tom Evans, Nikolay Nikolov, Jayesh Goyal, Jessica Spake, and Hannah Wakeford for reviewing the manuscript. DKS acknowledges funding from the European Research Council under the European Unions Seventh Framework Programme (FP7/2007-2013)/ERC grant agreement number 336792.

References

Bailey, J.: PASA **31**, 043 (2014)
Bean, J.L., Miller-Ricci Kempton, E., Homeier, D.: Nature **468**, 669 (2010)
Benneke, B., Seager, S.: ApJ **753**, 100 (2012)
Bétrémieux, Y., Swain, M.R.: MNRAS **467**, 2834 (2017)
Boss, A.P.: Science **276**, 1836 (1997)
Brown, T.M.: ApJ **553**, 1006 (2001)
Brown, T.M., Charbonneau, D., Gilliland, R.L., Noyes, R.W., Burrows, A.: ApJ **552**, 699 (2001)
Burrows, A., Budaj, J., Hubeny, I.: ApJ **678**, 1436 (2008)
Burrows, A., Hubeny, I., Budaj, J., Knutson, H.A., Charbonneau, D.: ApJ **668**, L171 (2007)
Burrows, A., Marley, M.S., Sharp, C.M.: ApJ **531**, 438 (2000)
Burrows, A.S.: Nature **513**, 345 (2014)
Carter, J.A., Winn, J.N.: ApJ **704**, 51 (2009)
Chandrasekhar, S.: Radiative transfer (1960)
Charbonneau, D., Brown, T.M., Latham, D.W., Mayor, M.: ApJ **529**, L45 (2000)
Charbonneau, D., Brown, T.M., Noyes, R.W., Gilliland, R.L.: ApJ **568**, 377 (2002)
Charbonneau, D., Allen, L.E., Megeath, S.T., et al.: ApJ **626**, 523 (2005)
Charbonneau, D., Berta, Z.K., Irwin, J., et al.: Nature **462**, 891 (2009)
Claret, A.: A&A **363**, 1081 (2000)
Cowan, N.B., Agol, E.: ApJ **729**, 54 (2011)
Croll, B., Lafreniere, D., Albert, L., et al.: AJ **141**, 30 (2011)
Crossfield, I.J.M.: PASP **127**, 941 (2015)
de Mooij, E.J.W., Snellen, I.A.G.: A&A **493**, L35 (2009)
de Wit, J., Seager, S.: Science **342**, 1473 (2013)
de Wit, J., Wakeford, H.R., Gillon, M., et al.: Nature **537**, 69 (2016)
Deming, D., Seager, S., Richardson, L.J., Harrington, J.: Nature **434**, 740 (2005)
Deming, D., Wilkins, A., McCullough, P., et al.: ApJ **774**, 95 (2013)
Deming, L.D., Seager, S.: JGRE **122**, 53 (2017)
Désert, J.-M., Lecavelier des Etangs, A., Hébrard, G., et al.: ApJ **699**, 478 (2009)
Diamond-Lowe, H., Stevenson, K.B., Bean, J.L., Line, M.R., Fortney, J.J.: ApJ **796**, 66 (2014)
Dragomir, D., Matthews, J.M., Eastman, J.D., et al.: ApJ **772**, L2 (2013)
Eastman, J., Gaudi, B.S., Agol, E.: PASP **125**, 83 (2013)
Eastman, J., Siverd, R., Gaudi, B.S.: PASP **122**, 935 (2010)
Espinoza, N., Fortney, J.J., Miguel, Y., Thorngren, D., Murray-Clay, R.: ApJ **838**, L9 (2017)
Evans, T.M., Aigrain, S., Gibson, N., et al.: MNRAS **451**, 680 (2015)
Evans, T.M., Sing, D.K., Kataria, T., et al.: Nature **548**, 58 (2017)
Fletcher, L.N., Orton, G.S., Teanby, N.A., Irwin, P.G.J., Bjoraker, G.L.: Icarus **199**, 351 (2009)
Foreman-Mackey, D., Hogg, D.W., Lang, D., Goodman, J.: PASP **125**, 306 (2013)
Fortney, J.J.: MNRAS **364**, 649 (2005)
Fortney, J.J., Lodders, K., Marley, M.S., Freedman, R.S.: ApJ **678**, 1419 (2008)
Fortney, J.J., Marley, M.S., Lodders, K., Saumon, D., Freedman, R.: ApJ **627**, L69 (2005)

Fortney, J.J., Mordasini, C., Nettelmann, N., et al.: ApJ **775**, 80 (2013)
Fortney, J.J., Shabram, M., Showman, A.P., et al.: ApJ **709**, 1396 (2010)
Gibson, N.P., Aigrain, S., Roberts, S., et al.: MNRAS **419**, 2683 (2012a)
Gibson, N.P., Nikolov, N., Sing, D.K., et al.: MNRAS **467**, 4591 (2017)
Gibson, N.P., Aigrain, S., Pont, F., et al.: MNRAS **422**, 753 (2012b)
Gillett, F.C., Low, F.J., Stein, W.A.: ApJ **157**, 925 (1969)
Gillon, M., Triaud, A.H.M.J., Demory, B.-O., et al.: Nature **542**, 456 (2017)
Grillmair, C.J., Burrows, A., Charbonneau, D., et al.: Nature **456**, 767 (2008)
Harrington, J., Hansen, B.M., Luszcz, S.H., et al.: Science **314**, 623 (2006)
Haswell, C.A.: Transiting Exoplanets (2010)
Hayek, W., Sing, D., Pont, F., Asplund, M.: A&A **539**, A102 (2012)
Haynes, K., Mandell, A.M., Madhusudhan, N., Deming, D., Knutson, H.: ApJ **806**, 146 (2015)
Helling, C., Woitke, P., Thi, W.: A&A **485**, 547 (2008)
Heng, K., Kitzmann, D.: MNRAS **470**, 2972 (2017)
Henry, G.W., Marcy, G.W., Butler, R.P., Vogt, S.S.: ApJ **529**, L41 (2000)
Hubbard, W.B., Fortney, J.J., Lunine, J.I., et al.: ApJ **560**, 413 (2001)
Hubeny, I., Burrows, A., Sudarsky, D.: ApJ **594**, 1011 (2003)
Huitson, C.M., Sing, D.K., Vidal-Madjar, A., et al.: MNRAS **422**, 2477 (2012)
Iro, N., Bézard, B., Guillot, T.: A&A **436**, 719 (2005)
Jensen, A.G., Redfield, S., Endl, M., et al.: ApJ **743**, 203 (2011)
Karkoschka, E., Tomasko, M.G.: Icarus **211**, 780 (2011)
Knutson, H.A., Charbonneau, D., Allen, L.E., Burrows, A., Megeath, S.T.: ApJ **673**, 526 (2008)
Knutson, H.A., Charbonneau, D., Noyes, R.W., Brown, T.M., Gilliland, R.L.: ApJ **655**, 564 (2007a)
Knutson, H.A., Charbonneau, D., Allen, L.E., et al.: Nature **447**, 183 (2007b)
Knutson, H.A., Dragomir, D., Kreidberg, L., et al.: ApJ **794**, 155 (2014)
Konacki, M., Torres, G., Jha, S., Sasselov, D.D.: Nature **421**, 507 (2003)
Kreidberg, L.: PASP **127**, 1161 (2015)
Kreidberg, L., Bean, J.L., Désert, J.-M., et al.: ApJ **793**, L27 (2014a)
Kreidberg, L., Bean, J.L., Désert, J.-M., et al.: Nature **505**, 69 (2014b)
Lecavelier des Etangs, A., Pont, F., Vidal-Madjar, A., Sing, D.: A&A, **481**, L83 (2008)
Lee, E.J., Chiang, E.: ApJ **817**, 90 (2016)
Line, M.R., Stevenson, K.B., Bean, J., et al.: AJ **152**, 203 (2016)
López-Morales, M., Seager, S.: ApJ **667**, L191 (2007)
Madhusudhan, N., Amin, M.A., Kennedy, G.M.: ApJ **794**, L12 (2014)
Magic, Z., Chiavassa, A., Collet, R., Asplund, M.: A&A **573**, A90 (2015)
Mandel, K., Agol, E.: ApJ **580**, L171 (2002)
Markwardt, C.B.: in Bohlender, D.A., Durand, D., Dowler, P. (eds.) Astronomical Society of the Pacific Conference Series, vol. 411, p. 251 (2009)
Marley, M.S., Ackerman, A.S., Cuzzi, J.N., Kitzmann, D.: in Mackwell, S.J., Simon-Miller, A.A., Harder, J.W., Bullock, M.A. (eds.) Clouds and Hazes in Exoplanet Atmospheres, p. 367 (2013)
Marley, M.S., Gelino, C., Stephens, D., Lunine, J.I., Freedman, R.: ApJ **513**, 879 (1999)
Mazeh, T., Naef, D., Torres, G., et al.: ApJ **532**, L55 (2000)
Morales-Calderón, M., Stauffer, J.R., Kirkpatrick, J.D., et al.: ApJ **653**, 1454 (2006)
Mordasini, C., Alibert, Y., Benz, W., Klahr, H., Henning, T.: A&A **541**, A97 (2012)
Morley, C.V., Fortney, J.J., Marley, M.S., et al.: ApJ **756**, 172 (2012)
Nikolov, N., Sing, D.K., Gibson, N.P., et al.: ApJ **832**, 191 (2016)
Nikolov, N., Sing, D.K., Burrows, A.S., et al.: MNRAS **447**, 463 (2015)
Nugroho, S.K., Kawahara, H., Masuda, K., et al. (2017). arXiv:1710.05276 [astro-ph.EP]
Öberg, K.I., Murray-Clay, R., Bergin, E.A.: ApJ **743**, L16 (2011)
Pollack, J.B., Hubickyj, O., Bodenheimer, P., et al.: Icarus **124**, 62 (1996)
Pont, F., Knutson, H., Gilliland, R.L., Moutou, C., Charbonneau, D.: MNRAS **385**, 109 (2008)
Pont, F., Sing, D.K., Gibson, N.P., et al.: MNRAS **432**, 2917 (2013)
Pont, F., Zucker, S., Queloz, D.: MNRAS **373**, 231 (2006)

Redfield, S., Endl, M., Cochran, W.D., Koesterke, L.: ApJ **673**, L87 (2008)
Richardson, L.J., Harrington, J., Seager, S., Deming, D.: ApJ **649**, 1043 (2006)
Ridgway, S.T.: ApJ **187**, L41 (1974)
Seager, S., Deming, D.: ARA&A **48**, 631 (2010)
Seager, S., Sasselov, D.D.: ApJ **537**, 916 (2000)
Showman, A.P., Guillot, T.: A&A **385**, 166 (2002)
Sing, D.K., López-Morales, M.: A&A **493**, L31 (2009)
Sing, D.K., Vidal-Madjar, A., Désert, J.-M., Lecavelier des Etangs, A., Ballester, G.: ApJ **686**, 658 (2008)
Sing, D.K., Pont, F., Aigrain, S., et al.: MNRAS **416**, 1443 (2011)
Sing, D.K., Wakeford, H.R., Showman, A.P., et al.: MNRAS **446**, 2428 (2015)
Sing, D.K., Fortney, J.J., Nikolov, N., et al.: Nature **529**, 59 (2016)
Snellen, I.A.G., Albrecht, S., de Mooij, E.J.W., Le Poole, R.S.: A&A **487**, 357 (2008)
Spiegel, D.S., Silverio, K., Burrows, A.: ApJ **699**, 1487 (2009)
Sromovsky, L.A., Fry, P.M., Kim, J.H.: Icarus **215**, 292 (2011)
Stevenson, K.B., Désert, J.-M., Line, M.R., et al.: Science **346**, 838 (2014)
Stevenson, K.B., Line, M.R., Bean, J.L., et al.: AJ **153**, 68 (2017)
Sudarsky, D., Burrows, A., Hubeny, I.: ApJ **588**, 1121 (2003)
Vidal-Madjar, A., Lecavelier des Etangs, A., Désert, J.-M., et al.: Nature **422**, 143 (2003)
Vidal-Madjar, A., Désert, J.-M., Lecavelier des Etangs, A., et al.: ApJ **604**, L69 (2004)
Vidal-Madjar, A., Sing, D.K., Etangs, Lecavelier Des, A., et al.: A&A **527**, A110 (2011)
Visscher, C., Lodders, K., Fegley Jr., B.: ApJ **716**, 1060 (2010)
Wakeford, H.R., Sing, D.K.: A&A **573**, A122 (2015)
Wakeford, H.R., Sing, D.K., Deming, D., et al.: MNRAS **435**, 3481 (2013)
Wakeford, H.R., Sing, D.K., Kataria, T., et al.: Science **356**, 628 (2017)
Wakeford, H.R., Sing, D.K., Deming, D., et al.: AJ **155**, 29 (2018)
Wallace, L., Prather, M., Belton, M.J.S.: ApJ **193**, 481 (1974)
Wong, M.H., Mahaffy, P.R., Atreya, S.K., Niemann, H.B., Owen, T.C.: Icarus **171**, 153 (2004)
Wyttenbach, A., Ehrenreich, D., Lovis, C., Udry, S., Pepe, F.: A&A **577**, A62 (2015)

Part II
Theoretical Models

Chapter 2
Modeling Exoplanetary Atmospheres: An Overview

Jonathan J. Fortney

Abstract We review several aspects of the calculation of exoplanet model atmospheres in the current era, with a focus on understanding the temperature-pressure profiles of atmospheres and their emitted spectra. Most of the focus is on gas giant planets, both under strong stellar irradiation and in isolation. The roles of stellar irradiation, metallicity, surface gravity, C/O ratio, interior fluxes, and cloud opacity are discussed. Connections are made to the well-studied atmospheres of brown dwarfs as well as sub-Neptunes and terrestrial planets, where appropriate. Illustrative examples of model atmosphere retrievals on a thermal emission spectrum are given and connections are made between atmospheric abundances and the predictions of planet formation models.

2.1 Why Study Atmospheres?

While atmospheres often make up only a tiny fraction of a planet's mass, they have an out-sized importance in determining a number of physical properties of planets, how they evolve with time, and their physical appearance. Atmospheres dramatically influence a planet's energy balance, as the relative importance of gaseous absorption or scattering from clouds or gasses dictate a planet's albedo. Atmospheres can impact cooling, as interior convection or conduction must give way to a radiative atmosphere to lose energy out to space. Atmospheres, by their composition, can tell us a rich story of the gain and loss of volatiles, since atmospheres can be accreted from the nebula, outgassed from the interior, lost to space by escape processes, or regained by the interior.

We tend to think of two broad reasons for studying planetary atmospheres. One is that atmospheres are inherently interesting, with a diverse array of physical and chemical processes at work. What sets the temperature structure of an atmosphere? Why do some have thermal inversions and others do not? What sets the chemical

J. J. Fortney (✉)
Other Worlds Laboratory (OWL), Department of Astronomy and Astrophysics, University of California, Santa Cruz, CA, USA
e-mail: jfortney@ucsc.edu

V. Bozza et al. (eds.), *Astrophysics of Exoplanetary Atmospheres*, Astrophysics and Space Science Library 450, https://doi.org/10.1007/978-3-319-89701-1_2

abundances in an atmosphere? Why are some atmospheres dominated by clouds, and why are others mostly cloud-free? What determines the day-night temperature contrast on the planet? How fast can winds blow? Can planets lose their entire atmosphere, never to attain one again?

An entirely other set of questions focuses more on what an atmosphere can tell us about the formation and evolution of the planet. Atmospheric composition can tell us a lot about the integrated history of a planet. The metal-enrichment of a giant planet, compared to its star, can help us to understand aspects of planet formation. The comparative planetology of rocky worlds, like Earth and Venus, one with water vapor in the atmosphere, and one without, informs our understanding of divergent evolution. Noble gas abundances teach us about the accretion of primordial volatiles.

The tools we use to model exoplanetary atmospheres are often the very same tools, or descendents of the tools, that we used to model the atmospheres of solar system planets. Other such tools were used to study cool stellar atmospheres or brown dwarfs. In that way, exoplanetary atmospheres can be thought of as a meeting of the minds, tools, and prejudice of the models and methods of planetary atmospheres and stellar atmospheres. The continuum from the coolest stars, to brown dwarfs and hot planets, to cool planets is real, and can be readily seen in Fig. 2.1.

The field of exoplanetary atmospheres has exploded in the past decade. With the 2020 launch of the *James Webb Space Telescope*, the field is poised for dramatic advances. We are lucky that a number of recent texts have emerged that discuss the

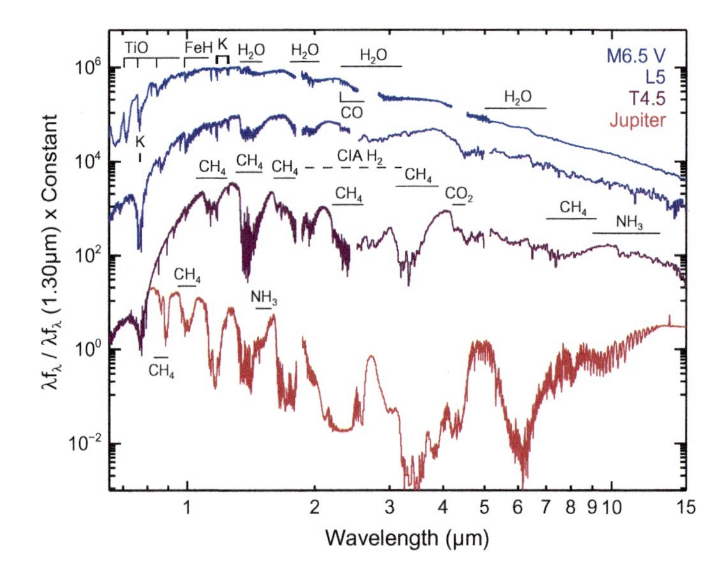

Fig. 2.1 An entirely empirical showcase of the continuum in atmospheres from the coolest stars, to brown dwarfs, to the spectrum of Jupiter. Dominant infrared molecular absorption features of H_2O, CO, CO_2, CH_4, and NH_3 are shown, as well as TiO and K at optical wavelengths. This sequence is T_{eff} values of 2700 K, 1550 K, 1075 K, and 125 K. Note that nearly all flux from Jupiter short-ward of 4.5 μm is reflected solar flux, not thermal emission. Figure courtesy of Mike Cushing

physics and chemistry of exoplanetary and solar system atmospheres. All are worth a detailed reading, including Seager (2010), Pierrehumbert (2010), Heng (2017), and Catling and Kasting (2017), while classic solar system texts like Chamberlain and Hunten (1987) are also still essential reading.

2.2 Energy Balance and Albedos

For any planet with an atmosphere, the atmosphere will help to set the energy balance of the planet with that of its parent star. Let's begin by striving to be clear about the albedo (reflectivity) of a planet, and how that enables estimates of planetary fluxes and temperatures. Often the descriptions of various albedos are actually much clearer in words than in mathematics, which is somewhat unusual. Excellent references on this topic exist from the "early days" of exoplanetary atmospheres, including Marley et al. (1999) and Sudarsky et al. (2000).

2.2.1 *Geometric Albedo*

Geometric albedo is the reflectivity of the planet when seen at full illumination, called "at full phase." (Think of the full moon.) Within the solar system, this is basically how we always see the giant planets, since they are on orbits at much larger separations than the Earth. In the exoplanet context, we can determine the geometric albedo of a planet at secondary eclipse (or "occultation") when its flux disappears as it passes behind its parent star. The geometric albedo, A_G, is always specified at a particular wavelength or in a given bandpass.

An oddity of A_G is that it can sometimes *fail* as a true measure of planetary reflectivity in the exoplanet context for hot planets (e.g., Burrows et al. 2008; Fortney et al. 2008). This is because hot Jupiters can have appreciable thermal emission at visible wavelengths. Thus, A_G can be higher than naively expected, due to some (or even most!) of the "planetary" flux coming from the planet being due to thermal emission, rather than reflected light.

2.2.2 *Spherical Albedo*

The spherical albedo, A_S, is similar to the geometric albedo. Again, it is specified at a given wavelength or a bandpass. However, here we are interested in the reflectivity over all angles—just the total reflectivity of the stellar flux, not caring about scattering angle. Recall that A_G is the stellar light that we get back when viewing the planet at full phase.

This means that A_S can only be determined with some care. Within the solar system, the most straightforward way is to send a spacecraft to the planet to observe scattering from the planet at all phase angles. In practice, the spherical albedo is not discussed much in the literature. However, if the spherical albedo is integrated over all *stellar* wavelengths, then we have a rather interesting quantity: the Bond albedo.

2.2.3 Bond Albedo

The Bond albedo, A_B, is typically the most important albedo for planets. It is the ratio of the total reflected stellar power (in say, erg s^{-1}) to the total power incident upon the planet. Within a modeling framework, it is A_S integrated over the stellar spectrum. The value of A_B is important because it determines how much total power is absorbed or scattered by a planet.

The single most important thing to recall about A_B is that, unlike A_G and A_S, *it is not a quantity that is inherent to the planet alone.* The value of A_B, for a given planet, *strongly depends on the incident spectrum from the parent star.* Meaning, the same planet, around two different stars, will have two different Bond albedos. Typically, around an M star, more flux is emitted in the infrared. There is less scattering, more absorption, and lower A_B, compared to illumination by a Sunlike star, where there is more short-wavelength incident flux that is Rayleigh scattered away (e.g., Marley et al. 1999).

While A_B is straightforward to discuss, it is difficult to measure in practice. Within the solar system, it can be determined by observing light scattered from planetary atmospheres in all directions (A_S) over a broad wavelength range that samples from the near UV to mid IR, where the Sun is brightest. In the exoplanet context, such a measurement is much more difficult. At least for strongly irradiated planets, A_B is probably best determined by just observing how hot a planet actually is, by measuring its total thermal emission.

2.2.4 Temperatures of Interest

Planets that do not have an intrinsic energy source will be in energy balance with the input from their parent star. That is, the power absorbed by the planet will be re-radiated back to space. For a planet like the Earth, the intrinsic energy due to secular cooling of the interior, along with radiative decay, is negligible in terms of energy balance; thus, absorbed power from the Sun entirely dominates the atmospheric energy balance. However, for very young rocky planets (Lupu et al. 2014), and for giant planets at essentially any age (Burrows et al. 1997; Baraffe et al. 2003; Marley et al. 2007), the flux from the planet's interior is appreciable and affects the atmospheric temperature structure and energy balance. If a planet is in energy balance with its star, the equilibrium temperature, $T_{\rm eq}$, can be written:

$$T_{\rm eq}^4 = f(1 - A_{\rm B})L_*/(16\pi\sigma d^2), \tag{2.1}$$

where f is 1 if the absorbed radiation is able to be radiated away over the entire planet (4π steradians) or 2 if it only radiates on the dayside (2π sr), which is then hotter. $A_{\rm B}$ is the planet's Bond albedo, L_* is the luminosity of the star, σ is the Stefan-Boltzmann constant, and d is the planet's orbital distance.

If the effective temperature, $T_{\rm eff}$, is defined as the temperature of a blackbody of the same radius that would emit the equivalent flux as the real planet, $T_{\rm eff}$ and $T_{\rm eq}$ can be simply related. This relation requires the inclusion of a third temperature, $T_{\rm int}$, the "intrinsic effective temperature," that describes the flux from the planet's interior. These temperatures are related by:

$$T_{\rm eff}^4 = T_{\rm eq}^4 + T_{\rm int}^4 \tag{2.2}$$

We then recover our limiting cases: if a planet is self-luminous (like a young giant planet) and far from its parent star, $T_{\rm eff} \approx T_{\rm int}$; for most rocky planets, or any planets under extreme stellar irradiation, $T_{\rm eff} \approx T_{\rm eq}$.

2.2.5 *Absorption and Emission of Flux*

Recent reviews on radiative transfer in substellar and exoplanetary atmospheres can be found in Hubeny (2017) and Heng and Marley (2017). Here will be merely show some illustrative plots of how a 1D radiative-convective atmosphere model operates in terms of the absorption of stellar flux (Fig. 2.2), the emission of thermal flux (Fig. 2.3), and the outgoing flux carried by an atmosphere in radiative-convective equilibrium (Fig. 2.4).

As we will see below, strongly irradiated planets are dominated by the absorbed incident stellar flux, rather than any intrinsic flux from the deep interior. Atmospheric energy balance is satisfied by re-radiation of absorbed stellar flux. For a generic gas giant planet at 0.05 au from the Sun, Fig. 2.2 shows the wavelength-dependent absorption of stellar flux at five pressure levels within a 1D model. At the top of the model, most flux is absorbed by the near infrared water bands (see Fig. 2.6 for their exact locations in wavelength), while at deeper layers most absorption is via the pressure-broadened alkali (Na and K) doublets at 0.59 and 0.77 μm. By 4 bars essentially all stellar flux has been absorbed. The re-radiation of this energy to space, at these same pressure levels, occurs in the near and mid infrared, mostly longward of 3 μm in the top three panels shown in Fig. 2.3. Deeper in the atmosphere, where temperatures are warmer, there is more overlap with shorter wavelength water bands.

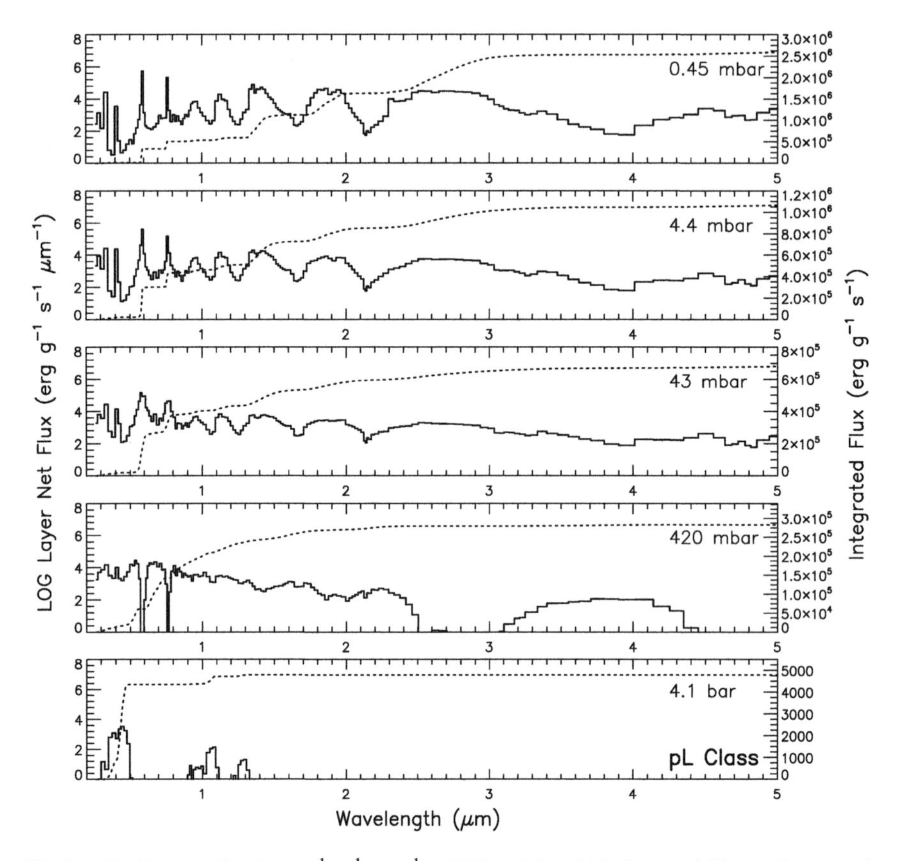

Fig. 2.2 Incident net flux (erg g^{-1} s^{-1} μm^{-1}, solid line, left axis) in five model layers for a cloud-free hot Jupiter model, $g = 15$ m s^{-2}, at 0.05 au from the Sun. The dotted line (right axis) illustrates the integrated flux, evaluated from short to long wavelengths. The layer *integrated* flux is read at the intersection of the dotted line and the right axis. Note the logarithmic scale on the left. At 0.45 mbar, although the absorption due to neutral atomic alkalis are important in the optical, more flux is absorbed by water vapor in the infrared. Heating due to alkali absorption becomes relatively more important as pressure increases. By 420 mbar, there is no flux left in the alkali line cores, and by 4.1 bar, nearly all the incident stellar flux has been absorbed. Adapted from Fortney et al. (2008), which also included a description of the hot Jupiter "pL Class" noted on the figure

Figure 2.4, adapted from Marley and Robinson (2015) shows the balance of several fluxes for a modestly irradiated planet, somewhat similar to Jupiter. The planet's intrinsic flux is carried via radiation or convection, with convection dominating at the deepest levels where the atmosphere is dense and mostly opaque. A second detached, convective zone forms in region of local high opacity, which carries some of the flux as well. At depth, the profile in the convective region (thicker solid line) is that of an isentrope (constant specific entropy) with a temperature-pressure profile that is

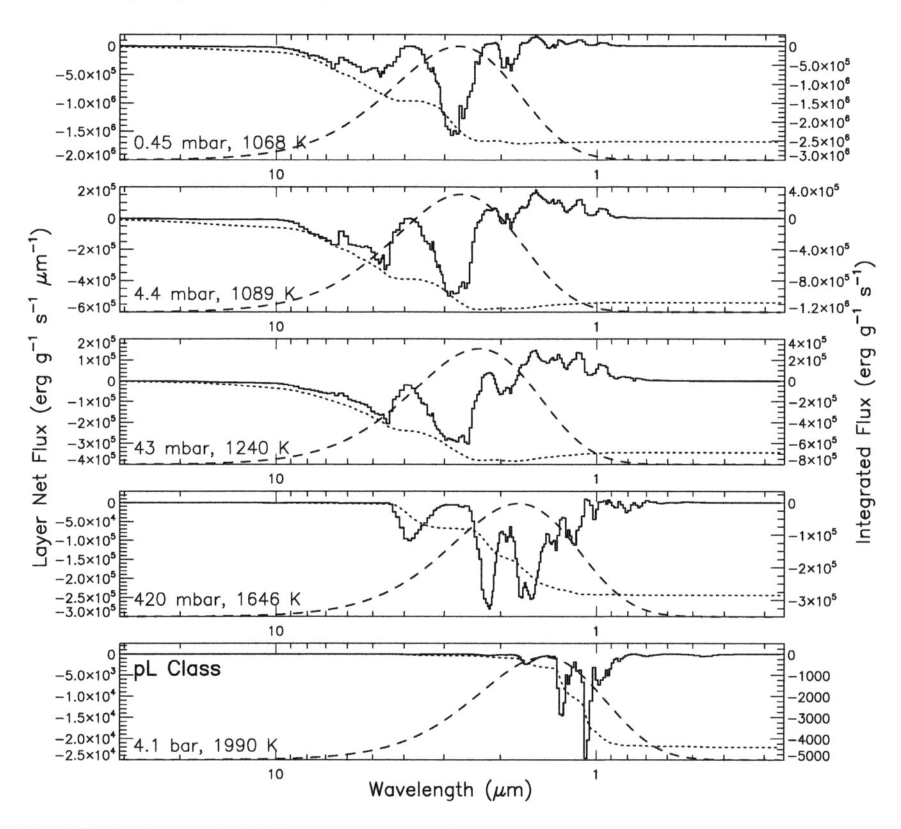

Fig. 2.3 Similar to Fig. 2.2, but for thermal emission, from 30 to 0.26 μm (long wavelengths to short). Negative flux is emitted. The dotted line is again integrated flux, evaluated from long to short wavelengths. The negative value of a layer integrated flux in Fig. 2.2 equals the integrated flux here. In addition, the scaled Planck function appropriate for each layer is shown as a dashed curve. The temperature and pressure of each layer is labeled. Cooling occurs mostly by way of water vapor, but also CO. As the atmosphere cools with altitude progressively longer wavelength water bands dominate the layer thermal emission. Adapted from Fortney et al. (2008)

essentially adiabatic, as a only a minute super-adiabaticity is needed to transport flux via convection. Absorbed stellar fluxes are shown as a an additional component that the atmosphere must also carry via radiation. In practice, a 1D radiative-convective model needs to iterate to find a temperature structure that satisfies the constant flow of intrinsic energy through each layer (given that each layer both absorbs and emits flux) and the re-emission of absorbed stellar flux at each layer. Marley and Robinson (2015) features an in-depth discussion of the temperature corrections needed in each model layer, for each iteration, to converge to a model in radiative-convective equilibrium.

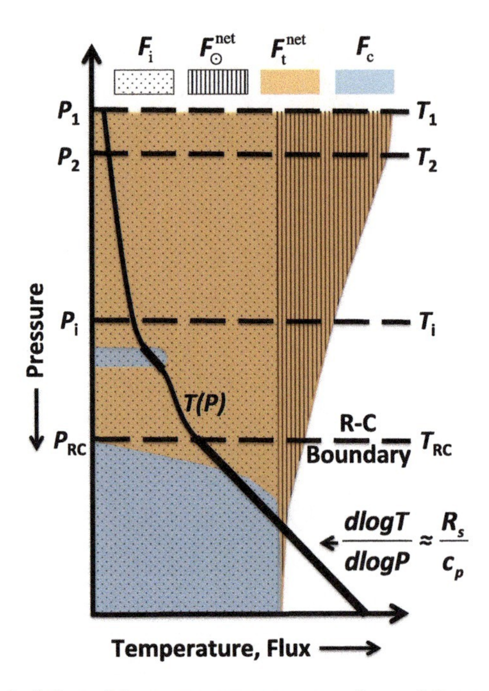

Fig. 2.4 Schematic depiction of the temperature structure of a model atmosphere. The y-axis is pressure, increasing downwards, and the x-axis shows temperature and energy flux. Model levels are shown (horizontal dashed lines), and the solid line is the temperature structure profile, where bolded parts indicate a convective region. 'RC' indicates the radiative-convective boundary. In equilibrium, net thermal flux (F_t^{net}, orange) and the convective flux (F_c, blue) must sum to the internal heat flux (F_i, dotted, which is σT_{int}^4) and, for an irradiated object, the net absorbed stellar flux ($F_\odot^{net}$, striped, which alone is σT_{eq}^4). Note that the internal heat flux is constant throughout the atmosphere, whereas the schematic profile of net absorbed stellar flux decreases with increasing pressure, and eventually reaches zero in the deep atmosphere. At depth, convection carries the vast majority of the summed internal and stellar fluxes, but is a smaller component in detached convective regions (upper blue region). Adapted from Marley and Robinson (2015)

2.3 Overview of Pressure-Temperature Profiles and Absorption Features

2.3.1 Pressure-Temperature Profiles

Much has been written over the years on the temperature structure of planetary atmospheres. Just in the recent past, analytic models of atmospheric temperature structure have been published, mostly focusing on strongly irradiated planets, by Hansen (2008), Guillot (2010), Robinson and Catling (2012), and Parmentier and Guillot (2014). These frameworks aim to understand the radiative (or radiative-convective) temperature structure as a function of the three temperatures outlined above, as well

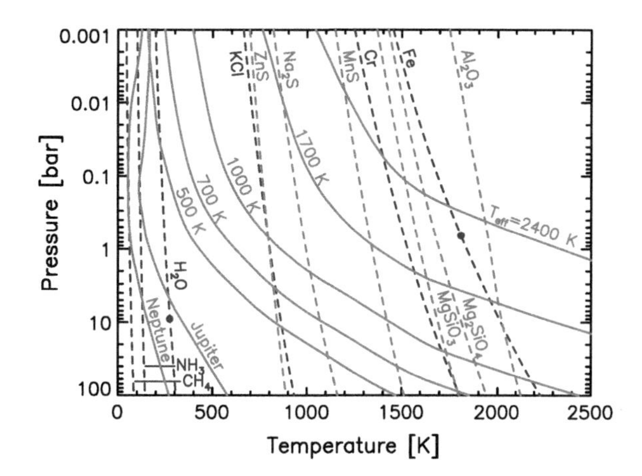

Fig. 2.5 Pressure-temperature profiles from a warm brown dwarf (2400 K) to Neptune (50 K), showing the range of cool molecule-dominated H-He atmospheres. Model atmospheric profiles are shown as solid curves. Chemical condensation curves for cloud species are shown as dashed lines. Figure adapted from Marley and Robinson (2015)

as the gaseous opacity relevant for flux incident upon the atmosphere (typically visible light) and the gaseous opacity relevant for emitted planetary fluxes (typically infrared light). Figure 2.5 shows atmospheric pressure-temperature (*T*–*P*) profiles from 2400 K down to 50 K, compared to relevant condensation curves for cloud-forming materials.

2.4 Interpreting Spectra via Absorption Features

While much of the physics of stellar atmospheres transfers over to our understanding of planetary atmospheres, the interpretation of spectra is not one such area. While much of stellar atmospheres can be interpreted in terms of narrow atomic/ionic lines, caused by electronic transitions, on top of wavelength-independent "continuum" opacity sources, the same is not true for planets. In planets, few—if any—continuum opacity sources exist, and atmospheric opacities are dominated by the forest of rotational/vibration lines of dominant molecular absorbers like H_2O, CH_4, CO, CO_2, and NH_3, among other molecules.

The entire concept of a "photosphere," the $\tau = 2/3$ surface from which all photons are emitted, is nearly meaningless in a planetary atmosphere, where opacity can vary widely from wavelength to wavelength. As an example, Fig. 2.6 shows the absorption cross-section (cm^2 per molecule) for a solar mixture of gases at 0.3 bar and three separate temperatures (2500, 1500, and 500 K) which shows that at all temperatures the opacity is nowhere dominated by any continua but instead by the opacities of various molecules.

The best way to interpret planetary emission spectra is with the concept of the brightness temperature, T_B. This is a wavelength-dependent quantity that is the temperature that a blackbody planet must have to emit the same amount of specific flux as the real planet, at that wavelength. For an atmosphere in local thermodynamic

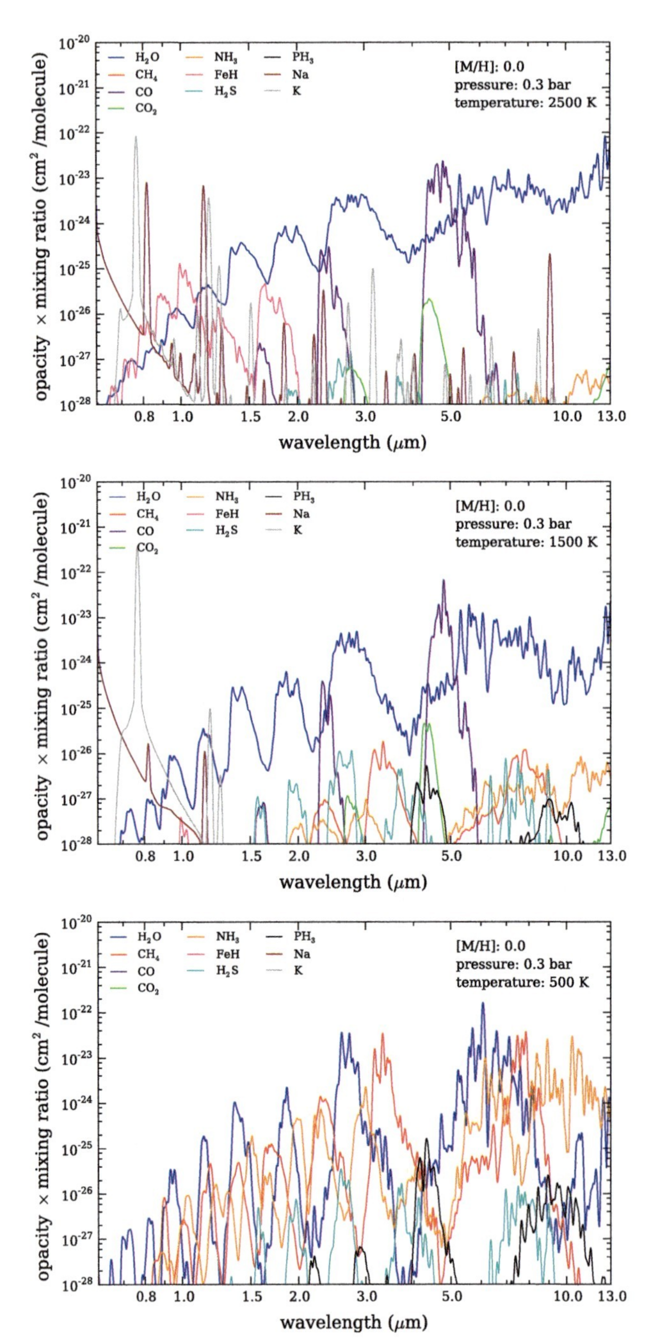

Fig. 2.6 These three panels show the absorption cross-sections of molecules, weighted by their volume mixing ratios. The x-axis is the wavelength range of interest for the *James Webb Space Telescope*. These calculations are for solar metallicity atmospheres at 0.3 bar, at 2500, 1500, and 500 K. Water vapor is a dominant opacity source at all of these temperatures. Figure courtesy of Caroline Morley

equilibrium (LTE), this corresponds to the temperature of the real atmosphere where the optical depth reaches 2/3—this is the level in the atmosphere that one "sees" down to. This is an important way to think about spectra as it can show us, for instance, that difference levels of flux in different infrared bands could actually come from the same level in the atmosphere, if they have the same T_B. The spectra and T_B, which are *both measured quantities*, can be turned into the *pressure level* of that thermal emission by interpolating the T_B values on a model pressure-temperature profile. One can then infer the wavelength-dependent pressure probed from an observed spectrum.

A more detailed analysis (e.g., Chamberlain and Hunten 1987) of emergent emission spectra shows, and one could likely intuit, that thermal emission at a particular wavelength comes from a range of pressures, not from only one precisely defined pressure. From this analysis emerges the definition of the "contribution function," which shows the *pressure range* from which thermal emission emerges. The pressure that corresponds to T_B is then merely the location of the maximum of the contribution function. The contribution function can be quantified, as by Knutson et al. (2009), as:

$$cf(P) = B(\lambda, T)\frac{de^{-\tau}}{d\log(P)} \tag{2.3}$$

A plot of the color-coded contribution function versus wavelength for a hot Jupiter atmosphere model is shown in Fig. 2.7.

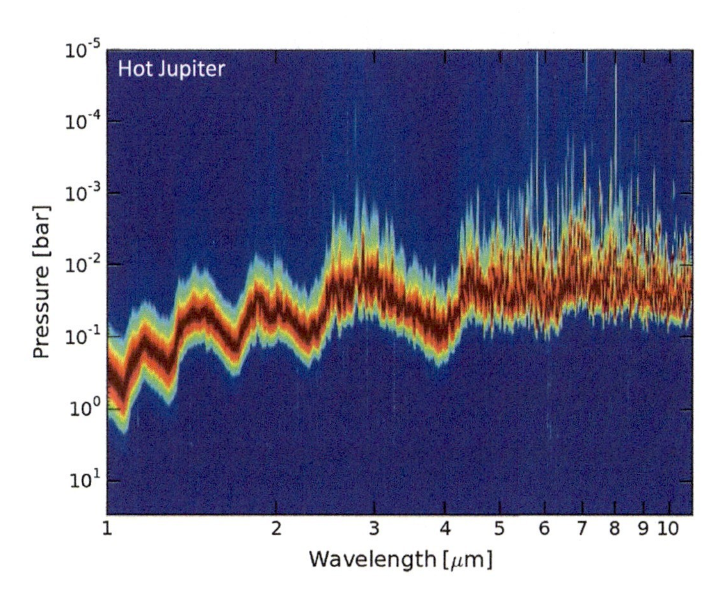

Fig. 2.7 Contribution function versus wavelength for an HD 209458b-like hot Jupiter. Maximum contributions are shown in red. Where opacity is highest, contributions come from the lowest pressures. At wavelengths of lower opacity, flux emerges from higher pressures. Figure courtesy of Mike Line

2.5 Stepping Through Physical Effects

In the following sections we will look at how a variety of physical and chemical processes affect the temperature-pressure profile and spectra of exoplanetary atmospheres. We will do this through a series of model calculations starting with, and then deviating from, solar-composition H/He atmospheres.

The atmosphere code employed for calculating these models iteratively solves for radiative-convective equilibrium by adjusting the size of the convection zone until the lapse rate everywhere in the radiative region is sub-adiabatic. This code was originally developed for modeling Titan's atmosphere (McKay et al. 1989), and has been extensively modified and applied to the study of brown dwarfs (Marley et al. 1996; Burrows et al. 1997; Saumon and Marley 2008; Morley et al. 2012) and solar and extrasolar giant planets (Marley et al. 1999, 2012; Fortney et al. 2005, 2008, 2013; Morley et al. 2015). The radiative transfer equations are computed using optimized algorithms described in Toon et al. (1989).

2.5.1 Surface Gravity

Typically, when one constructs a grid of model atmospheres, the first parameters of interest are the T_{eff} and the surface gravity. The surface gravity is often written as a log (in cgs units), such that "log $g = 4.0$" means a gravity of 10^4 cm s^{-2}, or 100 m s^{-2}. For reference, Jupiter's surface gravity is around 25 m s^{-2}. Surface gravity is a unit of choice because it is flexible. We may not know the masses and radii of the objects that we are studying so these quantities can be swept into the gravity.

All things being equal, lower gravity objects have emission from lower atmospheric pressures. This is clearly seen in Fig. 2.8, which shows the *T–P* profiles of two models with T_{eff} of 1000 K. The black dot on the profiles (top panel) at 1000 K show where the local temperature is equal to the model T_{eff}. This can be thought of as a kind of "mean photospheric pressure," although keep in mind that the emission comes from a range of pressures. Lower gravity leading to a lower photospheric pressure can be understood as the effect of gravity on the scale height, H, where $H = kT/\mu m_{\mathrm{H}} g$, and k is Boltzman's constant, T is the temperature, μ is the dimensionless mean molecular weight (2.35 for a solar abundance mix of gasses), m_{H} is the mass of the hydrogen atom, and g is the surface gravity. For an isothermal, constant gravity, and exponential atmosphere, the column density of molecules (N), from some reference location with local density n_o, vertically to infinity, is $N = n_o H$. This value N is directly proportional to the optical depth, τ. In either a low gravity or high gravity atmosphere, we see down to the $\tau = 2/3$ level at a given wavelength. If the gravity is lower at the same temperature, H will be larger, meaning that n_o will be smaller (found at a lower pressure) to reach the same N, or the $\tau = 2/3$ level.

We should expect the spectra of these two profiles to differ given their quite different temperature structures, and that is what we see in the second and third

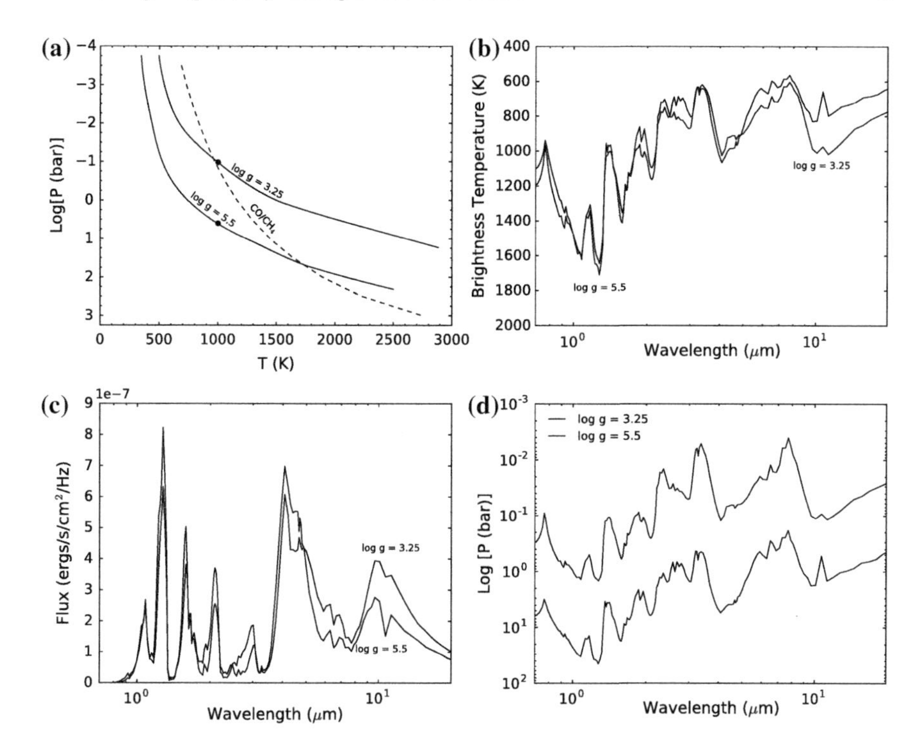

Fig. 2.8 These four panels demonstrate the influence of surface gravity. Shown in each are two models with the same T_{eff} of 1000 K, but surface gravities that differ by a factor of 178. Panel **a** shows the pressure-temperature profiles. Panel **b** shows the T_{B} for each wavelength. Panel **c** shows the emitted spectrum for both models. Panel **d** shows the pressure of the $\tau = 2/3$ layer as a function of wavelength, which graphically shows that lower-gravity atmospheres have lower pressure photospheres (see text). Pressure-dependent opacity like hydrogen collision-induced absorption (CIA) limits the depth one can see around 2 μm and longward of 10 μm in the higher gravity model

panels. A few things are worth noting: In the higher gravity atmosphere, the potassium doublet at 0.77 μm (first seen in Fig. 2.6 above) is much more pressure-broadened. Also, the flux peaks in the J (1.2 μm), H (1.6 μm), and K (2.2 μm) differ significantly. The high gravity model is much brighter in J and dimmer in K, compared to the low gravity model. This is due to the opacity source known a hydrogen "collision induced absorption" (CIA), which goes as the square of the local density. In higher pressure photospheres, this opacity source, which peaks in K band, is significantly more important. Most of the rest of the spectral differences can be attributed the differing abundances of CO and CH_4. The dashed curve shows where these molecules have an equal abundance in thermochemical equilibrium. To the right of this curve, CO is dominant, and to the left, CH_4 is. As one travels further from this curve, less and less of the "unfavored" species is found in the atmosphere.

2.5.2 Metallicity

The abundances of atoms and molecules in an atmosphere obviously also dictate the depth to which one can see at a given wavelength, and hence, the emitted spectra. For a H/He dominated atmosphere, as the metallicity increases, the opacity increases, and the photospheric pressures decrease. This can be seen in Fig. 2.9, which shows four models, all with $T_{\rm eff} = 1000$ K, but with metallicity values of [M/H]= −0.25, 0.0, +0.5, and +1.0. [M/H] is a log scale referenced to solar abundances, such that "0.0" is solar and "+1.0" is ten times solar. The metal-rich models have lower pressure photospheres so their deep atmospheres end on a warmer adiabat.

The spectra of these atmospheres for the most part look fairly similar. The main differences here are due to how metallicity influences chemical composition. The abundances of CO and CO_2 scale linearly and quadratically with metallicity, respectively (Lodders and Fegley 2002). This is seen most clearly from 4–5 μm (again refer to Fig. 2.6, where the metal-rich models show significantly more absorption from CO/CO_2).

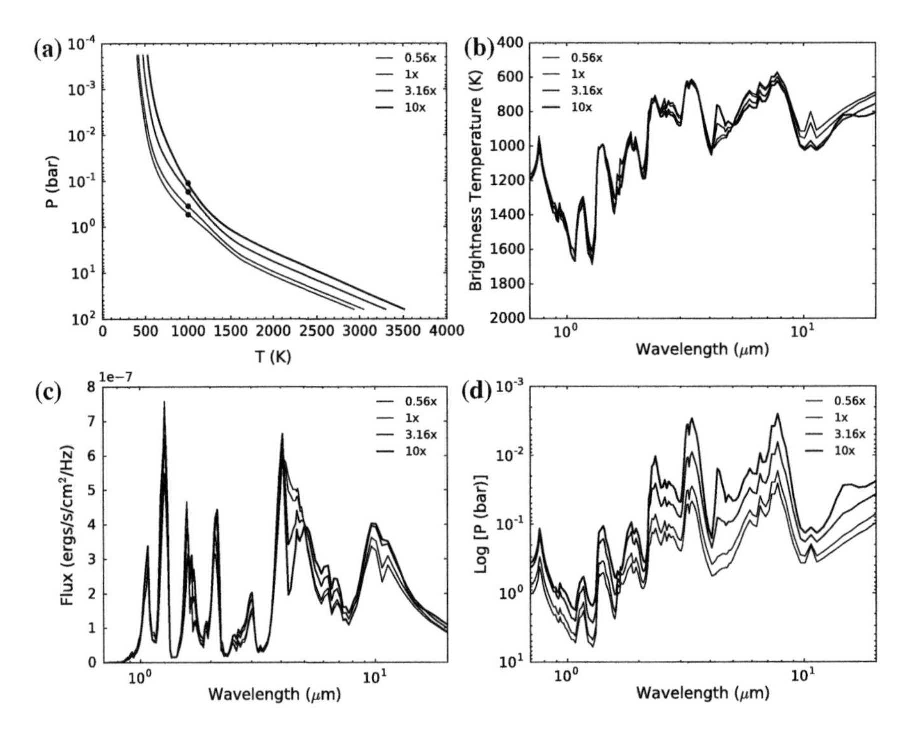

Fig. 2.9 Shown are four 1000 K log $g = 4$ models at four different metallicities, −0.25, 0.0 (solar), +0.5, and +1.0. The thickest line is the highest metallicity. Panel **a** shows the *P*–*T* profiles. Panel **b** shows the T_B values for these models. Panel **c** shows the emitted spectra. Panel **d** shows the wavelength-dependent pressure level of the $\tau = 2/3$ layer. The spectra of the models look fairly similar, except for much larger absorption due to CO and CO_2 in the metal-rich models, as CO and CO_2 and increase linearly and quadratically with metallicity, respectively

2.5.3 Carbon-to-Oxygen Ratio

In a solar metallicity gas, the carbon-to-oxygen (C/O) ratio is 0.54 (Asplund et al. 2009). In such a mixture, at higher temperatures, CO takes up just over half of the available oxygen, leaving the remainder of the oxygen to be found in H_2O. At cooler temperatures, carbon is found in CH_4, leaving most of the oxygen free to be in H_2O. This is readily seen in the spectral sequence of cool brown dwarfs, as seen in Fig. 2.1 and Kirkpatrick (2005). The C/O ratio also effects the condensation sequence of the elements which are lost into clouds (Fig. 2.5), as many of these refractory species take oxygen out of the gas phase and into the solid phase.

However, the details of particular abundances of molecules at a given P, T, and base elemental mixing ratios, quite sensitively depend on the abundances of C and O, as one might expect. In particular, there is a dramatic change at C/O $>$ 1. If C/O $>$ 1, then at hot temperatures nearly all oxygen will be tied up in CO, with little left for H_2O. Extra carbon can then go into CH_4, which is never seen at high temperatures for "normal" C/O ratios. The implications for spectra and the atmospheric structure of hot Jupiters have been examined in detail by Madhusudhan et al. (2011a, b), and Mollière et al. (2015).

Here we will examine the changes in the P–T profiles and spectra for three illustrative values of the C/O ratio in Fig. 2.10 for a 1600 K, log $g = 4.0$ model. These numbers are shown referenced to the solar value, such that C/O=1.0 is the solar value, *0.25* is 1/4 the solar value, and *2.5* is 5/2 of the solar value. These 1600 K models are hot enough that CO is the dominant carbon carrier. The models have relatively similar temperature structures, but the highest C/O value yields the highest pressure photosphere. This is because this model has the most C and O tied up in CO gas, which is relatively transparent in the infrared compared to H_2O and CH_4 (see Fig. 2.6). The spectra of the solar model and the *0.25* model are relatively similar, as both are dominated by H_2O opacity, with some contributions from CO. However, for the high C/O model, we can see that the H_2O bands essentially vanish and are replaced by strong CH_4 bands, which dramatically alters the emitted spectrum. Such objects have not been seen in the collection of brown dwarfs, but it is possible there are formation pathways for such carbon-rich giant planets, as discussed in Sect. 2.8.

2.5.4 Incident Flux

The incident flux from the parent star, often known as irradiation, insolation, or instellation, has a dramatic effect on the temperature structure of a planet. Indeed, for a terrestrial planet, the incident flux is the planet's only important energy source. For the Earth, most of the Sun's flux penetrates the atmosphere, which is optically thin in most of the optical, and is absorbed or scattered at the surface. This provides the Earth's atmosphere with a warm bottom boundary. Earth's atmosphere is optically thick at thermal infrared wavelengths near the surface, such that convection dominates in our troposphere.

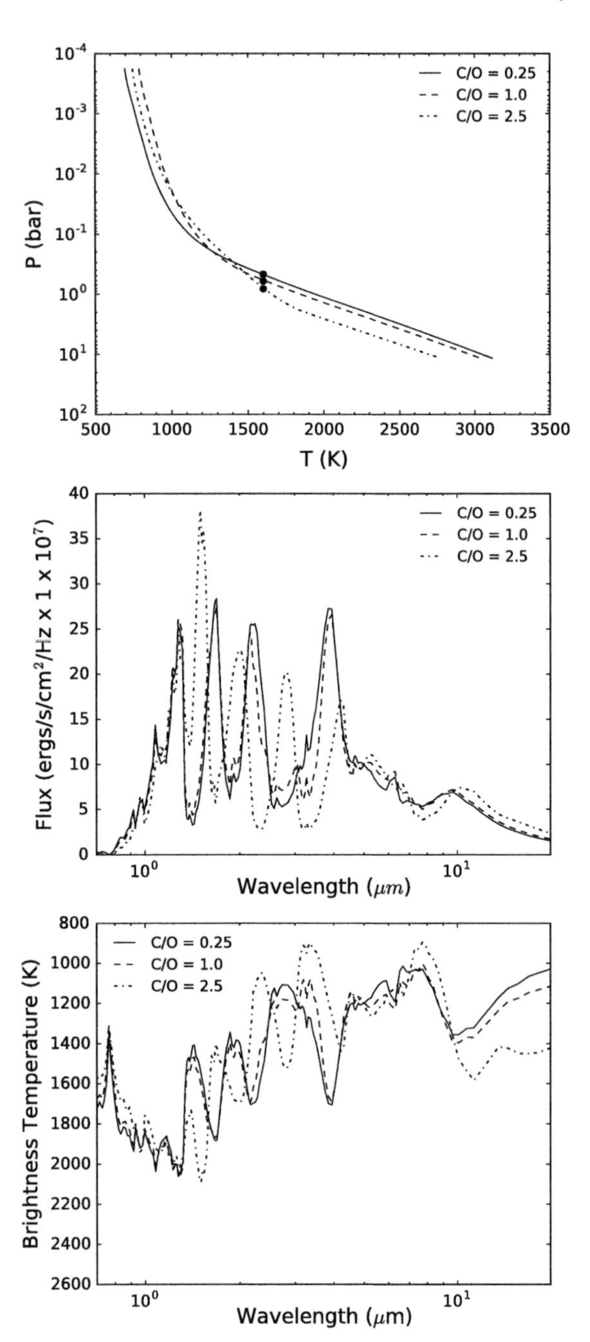

Fig. 2.10 These three panels show the effect of the atmospheric C/O ratio for three models at $T_{\rm eff} = 1600$ K and $\log g = 4$. Differences are mostly subtle in the P–T profile and are predominantly due to changes in the water vapor opacity, which is the dominant absorber

For a strongly irradiated planet, such as a hot Jupiter at 0.03 au, or even a sub-Neptune at 0.2 au, the absorption and re-radiation of stellar flux carves the planetary *T–P* profile shape. An illustrative example of the difference between a H/He atmosphere heated from below (like a young giant planet on a wide orbit, or brown dwarf) or a planet heated from above by stellar flux, is show in Fig. 2.11. These two models have the same $T_{\rm eff}$ of 1600 K and same surface gravity of ($\log g = 4$), but the temperature structures appear radically different. In this figure, the convective parts of the atmosphere are shown in a thicker line, while the radiative parts are shown as a thinner line. The irradiated model is forced to have a much hotter upper atmosphere. This also forces the more isothermal, radiative part of the atmosphere to be relatively large in vertical extent, pushing the radiative-convective boundary down to nearly 1000 bar. This is a generic finding for all strongly irradiated atmospheres and it is a significant difference between these atmospheres and those under modest stellar irradiation. Weakly irradiated atmospheres can typically be convective up to the visible atmosphere. Mixing processes in the radiative part of the atmosphere are probably slower than in the convective part of the atmosphere, but it is incorrect to think of these radiative regions as being quiescent.

The spectra of the two models (middle panel) look almost nothing alike, owing to the radically different temperature structures. In the near infrared, where one sees most deeply into the atmosphere (see Fig. 2.6, middle panel), the isolated model is significantly hotter, which yields much higher near-infrared fluxes. The bottom panel of Fig. 2.11 graphically shows that the more isothermal temperature structure directly leads to a smaller dynamic range in the temperatures probed, which leaves the spectrum more blackbody-like than the isolated object. This suggests that while brown dwarfs and imaged planets will provide (and have been providing) essential lessons in terms of atmospheric abundances and molecular opacities, we should *not* expect spectra of the strongly irradiated planets to necessarily follow the same sequence that is clear for the isolated objects at these same $T_{\rm eff}$ values (e.g., Kirkpatrick 2005). Figure 2.12 shows the result of a calculation of placing this same model planet at different distances from its parent star to yield $T_{\rm eff}$ values from 2400 to 600 K. All models have a $T_{\rm int}$ value of 200 K.

2.5.5 Outer Boundary Condition: Parent Star Spectral Type

As we have seen in Figs. 2.2 and 2.3, and the previous section, the pressure levels and wavelengths at which incident stellar flux is absorbed and then re-radiated back to space dictate the temperature structure, especially for strongly irradiated planets. Not all parent stars have the same spectra energy distributions, as hotter A-type parent stars will peak in the blue or even UV, while cool M-star hosts will peak in the near-infrared, in accordance with Wien's law.

It is difficult to create a one-size-fits-all grid of model atmospheres for strongly irradiated planets because each particular planet has its own particular parent star. Typically, when modeling a given exoplanet atmosphere, investigators will create

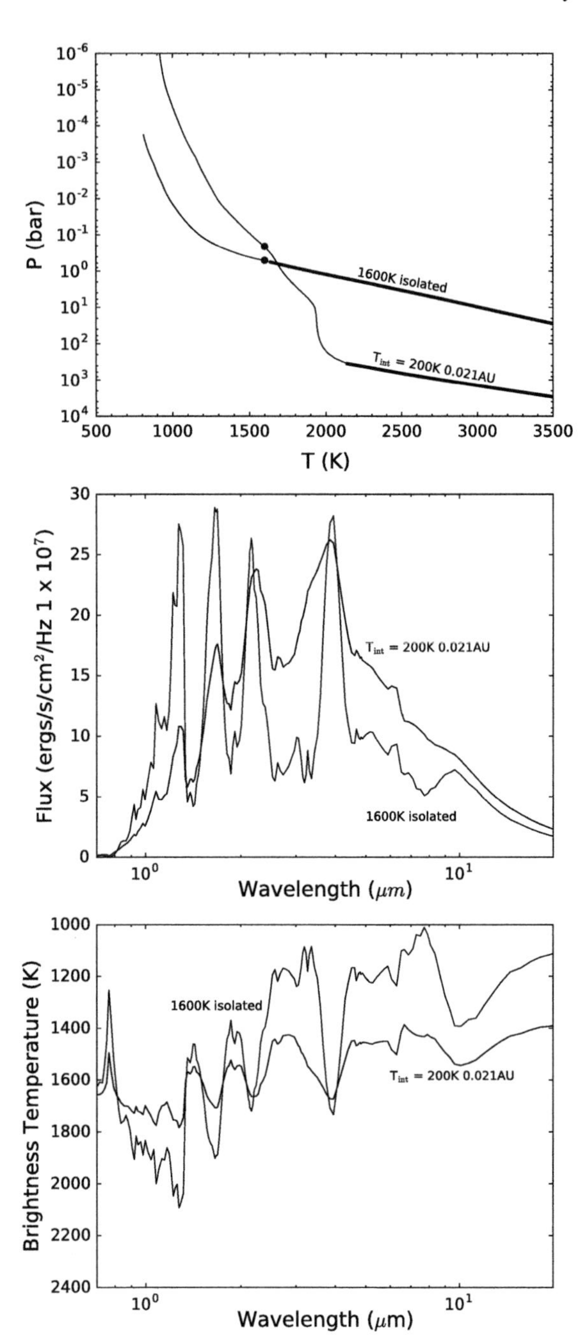

Fig. 2.11 Shown are two models with the same $T_{\rm eff}$(1600 K) and surface gravity ($\log g = 4$). One is an isolated object and one is in a close-in orbit around a main sequence G0 parent star. The irradiated planet model has a much shallower temperature gradient. The radiative-convective boundary is pushed to much higher pressures in the irradiated model. In the top panel, the convective part of the atmosphere is shown with a thicker line. The more isothermal atmosphere yields a more muted (modestly more blackbody-like) emission spectrum

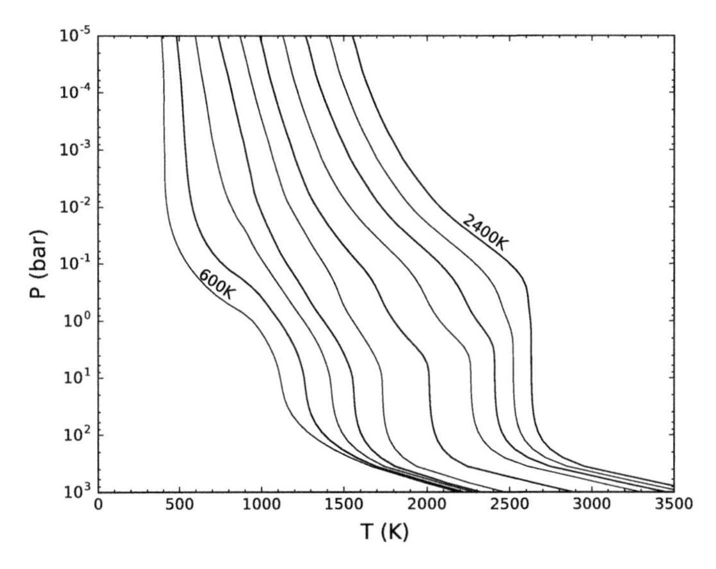

Fig. 2.12 A sample calculation of the effects of stellar insolation, for planets at various distances from a 6000 K G0 main sequence star. The models go from $T_{\rm eff}$ of 2400 to 600 K in 200 K increments. All models have a $T_{\rm int}$ value of 200 K

a synthetic spectrum for the parent star by interpolating in a grid of stellar model atmospheres (Hauschildt et al. 1999) for the fluxes incident upon the planet. This effect of stellar spectral type on hot Jupiter atmospheres has been investigated in some detail by Mollière et al. (2015).

Here in Fig. 2.13 we look at just a subset of models, for a 6000 K type G0 and 5000 K type K2 parent star. These models are placed at distances such that the planet $T_{\rm eff}$ are the same, yielding planetary $T_{\rm eff}$ values of 1000 and 1800 K. The cooler parent star spectrum (solid curves), peaking at longer wavelengths, allows more incident energy to be absorbed higher in the atmosphere by the water bands (see Fig. 2.2), which warms the upper atmosphere and cools the lower atmosphere, relative to the hotter parent star (dotted curves). The spectra from the more isothermal atmosphere are, as expected, slightly more muted, since the spectrum of a truly isothermal atmosphere would appear as a blackbody.

2.5.6 *Inner Boundary Condition: Flux from the Interior*

Typically, one worries little about the inner boundary condition for a strongly irradiated planet. Given the three temperatures discussed above, $T_{\rm eff}^4 = T_{\rm eq}^4 + T_{\rm int}^4$, the planetary $T_{\rm eff}$ is typically dominated by absorbed stellar flux, with $T_{\rm int}$ contributing little to the planetary energy budget. As one moves further from the parent star, or to younger planets which have interiors that have not yet cooled with age, the flux from

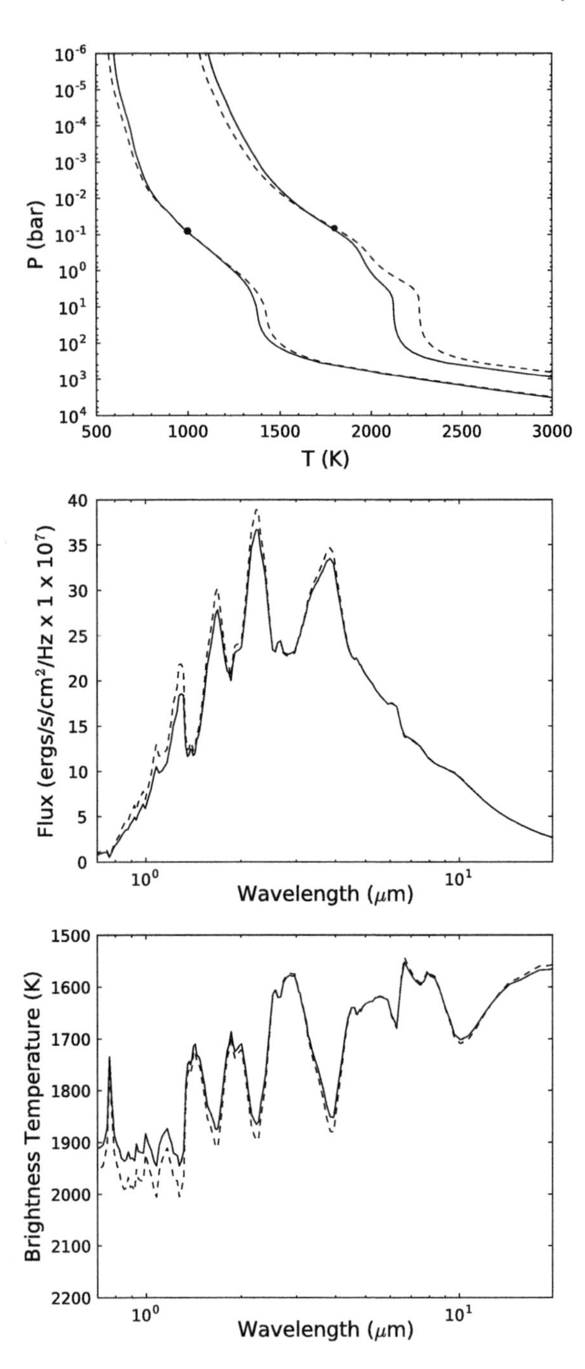

Fig. 2.13 The effect of parent star spectral type on two hot Jupiter atmosphere models at T_{eff} values of 1000 and 1800 K. The solid line is a main sequence K2 star, while the dashed line is G0. The cool star puts out more flux in the infrared, which can be absorbed higher in the atmosphere by strong infrared bands leading to a slightly shallower *P-T* profile, which mutes the spectrum. The spectrum is only shown for the hotter model

the interior can matter considerably. Obviously, for isolated brown dwarfs or young gas giants on wide separation orbits, this intrinsic flux is essentially all the flux that we see.

The inner boundary for strongly irradiated gas giants may be important in energy balance in some circumstances, however (Fortney et al., in prep). As shown in Morley et al. (2017), the emission spectrum of prototype warm Neptune GJ 436b is best matched by a model that has a high $T_{\rm int}$ value of $\sim$350 K, much higher than the $\sim$50 K one would expect from a Neptune-like evolution model (e.g., Fortney et al. 2011). Morley et al. (2017) suggest the large intrinsic flux is due to ongoing tidal dissipation, as the planet is currently on an eccentric orbit. With the aid of a tidal model these authors constrain the tidal Q of the planet. This was an interesting planetary science exercise where the planet's emission spectrum was tied to its interior structure and orbital evolution!

Another case where the inner boundary may matter is for the largest-radius ("most inflated") hot Jupiters. This is because large radii imply hot interiors, which implies high fluxes from the interior (e.g., Fortney et al. 2007). Thorngren and Fortney (2017) recently suggested that some hot Jupiter's may have $T_{\rm int}$ values of $\sim$700 K, far in excess of Jupiter's value of 100 K. Figure 2.14 shows an example calculation for a hot Jupiter at 0.03 au from the Sun, with inner boundaries $T_{\rm int}$ values from 100 to 700 K. The flux enhancements are most prominent in the JHK near-infrared windows, which probe deepest in the planetary atmosphere. Although small, these altered fluxes are likely detectable with *JWST*.

2.5.7 Role of Atmospheric Thickness

For gas giant planets the atmospheric thickness takes up most of the radius of the planet. Even for a sub-Neptune, we have little hope of seeing the bottom of the atmosphere. For example, Lopez and Fortney (2014) point out that for a 5 $M_\oplus$ rocky planet with only 0.5% of its mass in H/He (which would yield a radius of 2 $R_\oplus$), the surface pressure would be 20 kbar.

However, for terrestrial planets the atmospheric thickness is tremendously important, as it helps to set the surface pressure. Surface temperature tends to scale with surface pressure as well, due to the greenhouse effect. Our nearby example is Venus, which actually has a *lower* $T_{\rm eq}$ than the Earth, due to Venus's high Bond albedo. Venus's extremely high surface temperature is mostly due to its atmospheric pressure, which is 90 times larger than the Earth's. Determining the surface pressure is not necessarily straightforward. Perhaps the most straightforward way is if one could detect signatures from the ground in the planetary spectrum, in wavelengths where the atmospheric opacity is low, such that it could be optically thin to ground level.

Figure 2.15 shows models from Morley et al. (2017b) that examine the spectra of planet TRAPPIST-1b, the innermost planet in the TRAPPIST-1 system, around a very late M dwarf. These plots examine the role of surface pressure on emission and transmission, with pressures from 10^{-4} to 10^2 bar. It should be noted that these models

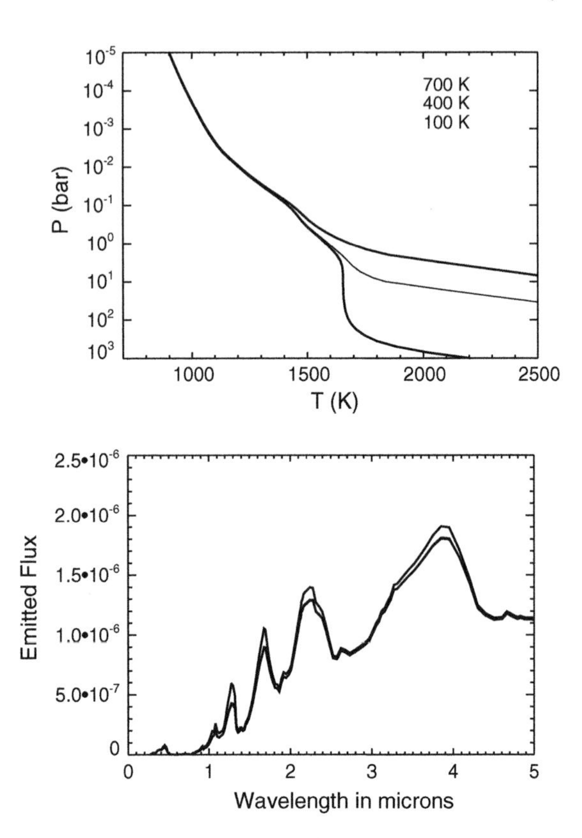

Fig. 2.14 The effect of a hotter and cooler inner boundary on the *T–P* profile and spectrum of a Saturn-like (gravity = 10 m s^{-2}) planet at 0.04 au from the Sun. Values of $T_{\rm int}$ are 700, 400, and 100 K, with higher values leader to a hotter deep atmosphere and shallower radiative-convective boundary. A hot inner boundary could be due to youth or additional energy sources in the planetary interiors. The spectra from the 400 and 100 K models are indistinguishable. The differences in spectra are modest but likely detectable with *JWST*

are cloud-free. In emission, in the top panel, the atmosphere is everywhere optically thin enough to see emission from the blackbody surface. In transmission, which has a much longer atmospheric path length (e.g., Fortney 2005), one can no longer see the surface for pressures above 10^{-3} bars. An interesting dichotomy between these plots is that for a wide range of thick atmospheres the transmission spectra are the same. However, the emission spectra differ substantially. This is because for transmission spectra, the atmosphere is basically a passive absorber of the stellar flux. In emission, flux originates from a diverse range of pressure levels, meaning that the atmospheres appear distinct until the pressure is high enough that the atmosphere is optically thick at all pressures, which is nearly seen in the emission spectra models at 10 and 100 bars.

2.5.8 Effects of Clouds

The phase change of molecules from the gas phase to the liquid or solid phase is an unavoidable consequence of lower temperatures. Within the solar system, clouds

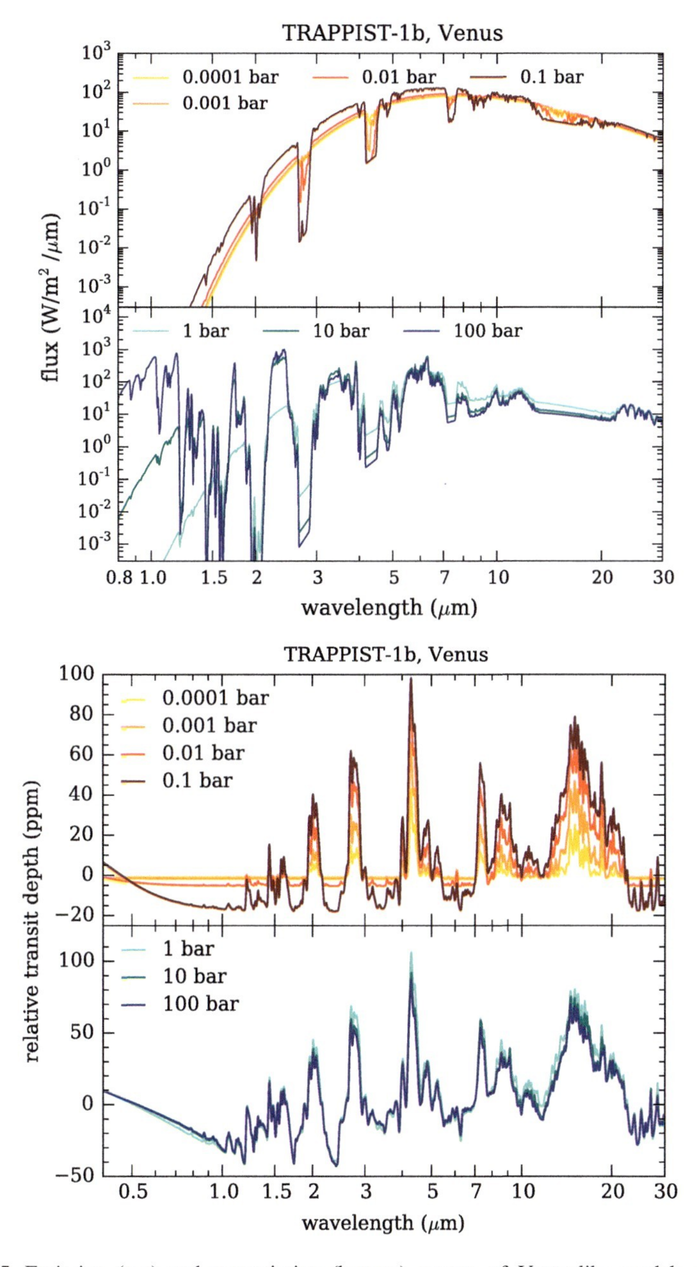

Fig. 2.15 Emission (top) and transmission (bottom) spectra of Venus-like models of planet TRAPPIST-1b, courtesy of Caroline Morley, from Morley et al. (2017b). Colors denote the surface pressure of the models. These models have an adiabatic P–T profile with depth that terminates at the planetary surface

are ubiquitous in every planetary atmosphere. As shown in Fig. 2.5, a wide range of species condense to form solids and liquids in H/He dominated atmospheres. While sometimes known as "dust" in the astrophysical literature, here we will reference these condensates as "clouds." At the highest temperatures, the most refractory species such as Al_2O_3 (corundum) and Ca-Ti-O bearing species will condense. Next are silicates ($MgSiO_3$ or Mg_2SiO_4) and iron. A variety of sulfide species condense at moderate temperatures, later followed by water (H_2O) and ammonia (NH_3) at the coolest temperatures. For reference, the cloud layers in Fig. 2.5 should all occur in Jupiter at great depth, as one could mentally extrapolate the deep atmosphere's adiabat to higher pressures, past $\sim$1 kbar or higher.

Many of the *P–T* curves that indicate condensation can be readily derived from the Claussius-Clapeyron relation, as discussed in Seager and Deming (2010). A detailed look at the chemistry of condensation across $\sim$500–1500 K can be found in Morley et al. (2012) and Morley et al. (2013), with applications to cool brown dwarfs and warm sub-Neptune transiting exoplanets, respectively. Cloud modeling is quite important because cloud opacity can be just as important as gaseous opacity. However, while gaseous opacity can in principle readily be measured over a range of P and T via laboratory work, or via first-principles quantum mechanical simulations (see the contribution from Jonathan Tennyson, Chap. 3), cloud opacity is generally not amenable to this kind of analysis.

The complexity of clouds comes from a number of reasons. Most importantly, there is a tremendous amount of complex and poorly understood *microphysics*. One must try to understand, at a given pressure level in the atmosphere, the mean or mode particle size, the (likely non-symmetric) distribution in sizes around this value, which could be bi-modal, how this mode and distribution change with height, the absorption and scattering properties, and scattering phase functions of these particles. All of these quantities likely change with latitude and longitude, as Earth and Jupiter both have clear and cloud patches. In addition, the coverage fraction of a given visible hemisphere likely changes with time.

There are several different ways investigators have aimed to understand the role of clouds in exoplanetary atmospheres. A relatively simple cloud model is that of Ackerman and Marley (2001) who aim to understand the 1D distribution of cloud particles by balancing the sedimentation of particles with the upward mixing of particles and condensible vapor. All microphysics is essentially ignored and parameterized by a sedimentation efficiency parameter, f_{sed} (called f_{rain} in the original paper). f_{sed} is tuned to fit observations of planets and brown dwarfs. This methodology has been beneficial and has been applied across a wide range of T_{eff}, from $\sim$200 to 2500 K. However, it lacks predictive power. Another framework is that of Helling and collaborators (Helling and Woitke 2006; Helling et al. 2008, 2017) who use a sophisticated chemical network and follow the microphysics of tiny "seed particles" that fall down through the planetary atmosphere, from the atmosphere's lowest pressures. Cloud particles sizes and distributions versus height naturally emerge from these calculations, which typically yield "dirty" grains of mixed compositions. The comparison of these models to observed planetary and brown dwarf spectra has not progressed quite as far at this time.

The effects of clouds on the spectra of planets can be readily understood. Cloud opacity is typically "gray," meaning that there is little wavelength dependence to the opacity. (While typical, this is not a rule.) Compared to a cloud-free model, clouds are an additional opacity source, and limit the depth to which one can see. Clouds typically then raise the planetary "photosphere" to lower pressures. In Fig. 2.16 we can examine the effect of clouds on a 1400 K, log $g = 4$ young giant planet. These examples lack external irradiation. These four panels show a cloud-free model (thin solid curve), an optically thinner cloud ($f_{\rm sed} = 4$) cloud as a dashed curve, and a thicker cloud ($f_{\rm sed} = 2$) in a thicker solid curve.

Figure 2.16a shows the atmospheric P–T profiles. The black dot shows where the local temperature matches the 1400 K $T_{\rm eff}$, which one can think of as the mean photosphere. With larger cloud opacity, the P–T profile is shifted to lower pressures at

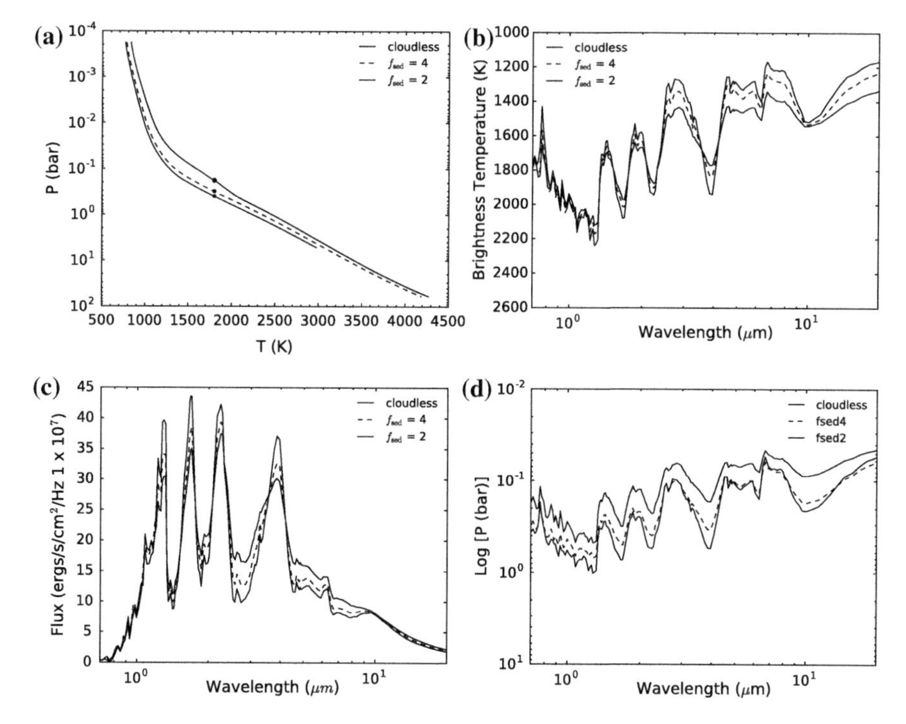

Fig. 2.16 The effect of clouds on a thermal emission spectrum of three models with the same $T_{\rm eff}$ (1800 K) and gravity (log $g = 4.0$). Panel **a** shows the atmospheric T–P profile, which is warmer with increased cloudy opacity, and has a lower pressure photosphere. This thinner solid line is cloud-free. Panel **b** shows the brightness temperatures one sees as a function of wavelength. Note how the effect of (gray) cloud opacity is to mute the "hills and valleys," and limit the depth to high temperatures one might see. This manifests in the spectrum in **c** where, with increasing cloud opacity, the spectrum becomes more muted and modestly more black-body like, and the JHK peaks are muted. Panel **d** shows the pressures probed (the $\tau = 2/3$ level) at each wavelength. Note the lower pressure photosphere as well as the smaller dynamic range in pressure as cloud thickness increases

a given temperature. Figure 2.16b shows the resulting brightness temperatures. In the near infrared the clouds limit our ability to see deeply down in the JHK near-infrared opacity windows. The inclusion of the gray opacity source limits the dynamic range in temperatures that are seen in the cloud-free model. For each profile, the pressure that corresponds to each value of the brightness temperature, where the optical depth out of the atmosphere is 2/3, is shown just below in Fig. 2.16d. The higher pressures probed in the cloud-free model, and the higher dynamic range of pressures probed, is clearly seen. The resulting emergent spectra are shown in Fig. 2.16c. Although all spectra are nothing like a blackbody, the cloudiest models have the most muted absorption features. The J-band at $\sim$1.3 μm, where water opacity is at its minimum, and one would normally see deepest into the atmosphere, is the wavelength range that is most affected by the cloud opacity.

After the past 15 years of observing transit transmission spectra, we are fully aware of how clouds manifest themselves in transiting planet atmospheres. Typically, weaker than expected absorption features are seen (e.g., Sing et al. 2016). There are numerous examples where cloud opacity blocks all molecular absorption features (Kreidberg et al. 2014b). An illustrative example of how clouds effect absorption features in transmission spectrum model is shown in Fig. 2.17. These models already

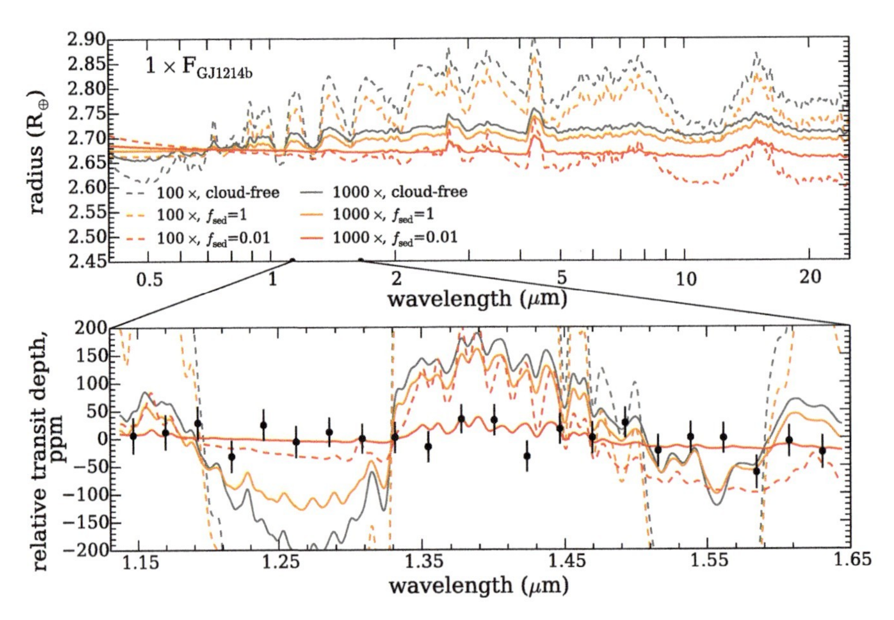

Fig. 2.17 Example transit transmission spectra of planet GJ 1214b at high metallicity, with and without clouds. Models are at metallicity values of 100 and 1000× solar. The top panel shows the optical and infrared transmission spectra. The bottom panel shows the same spectra, zoomed in to focus on the featureless Kreidberg et al. (2014b) data in the near-infrared from *Hubble*. Cloud-free transmission spectra are shown as dotted and solid gray lines and cloudy spectra are shown as colored lines. Note that the only model that fits the data is the 1000× solar model with $f_{sed} = 0.01$ (very highly lofted) clouds. Figure courtesy of Caroline Morley

have enhanced metallicity atmospheres that drive up the mean molecular weight, μ, which shrinks the scale height and size of absorption features. The clouds provide additional (mostly gray) opacity that mutes the absorption features.

2.6 Retrieval

Over the past several years inverse methods have become an integral part of modeling exoplanetary atmospheres. Within this framework one performs a wide exploration of a range of possible atmospheres models that can yield a best-fit to an observed spectrum. Typically these methods, called "retrieval," aim to find the combination of atmospheric abundances and atmospheric T–P profile that yield a best fit to an observed spectrum. Those models can find solutions outside the confines of self-consistent models.

2.6.1 Forward Model

The most important piece of any atmospheric retrieval algorithm is the *forward model*. The forward model takes a set of inputs, generally the parameters of interest, and then uses various physical assumptions to map these parameters onto the observable, e.g., the spectra. Depending on the situation of interest, there are three kinds of forward models one would implement, depending on if the observations were in thermal emission, transit transmission, or reflection. First is an emission forward model, which computes the upwelling top-of-atmosphere flux, and would be used at secondary eclipse or for directly imaged planets. Second is a transmission forward model, which computes the wavelength-dependent transit radius of the planet, and is used in defining the planet-to-star radius ratio. Third is a reflection forward model, which sums the stellar flux scattered in any direction from an illuminated hemisphere, and would be used as a function of orbital phase for an imaged or a transiting planet phase curve.

These forward models differ in their geometry and in the atmospheric regions probed. The necessary inputs in all models are the temperatures at each atmospheric level–the thermal profile–and the abundance and type of opacity sources, whether they be molecular/atomic gases or cloud/grain opacities, gravity, host star properties, and basic instrumental parameters that convolve and bin the high-resolution model spectra to the data set in question. These are the fundamental quantities that impact any emission, transmission, or reflection spectrum.

2.6.2 Bayesian Estimator/Model Selection

The Bayesian estimator is used to determine the allowed range of parameters (posterior), in the context of a given forward model, which can adequately describe the data. Most investigators use a multi-pronged modeling approach (e.g., Line et al. 2013) that includes several MCMC samplers, including the powerful ensemble sampler EMCEE (Foreman-Mackey et al. 2013), implemented in Kreidberg et al. (2015), Greene et al. (2016), and multimodal nested sampling (Feroz et al. 2009), PYMULTINEST (Buchner et al. 2014), implemented in Line et al. (2016). Using more than one inference method ensures robust results (e.g., Lupu et al. 2016).

An important aspect of Bayesian problems is that of model selection. Given two competing models, model selection aims to rigorously identify the simplest model that can adequately explain the data. This is done through the evaluation of the marginal likelihood, or Bayesian evidence (e.g., Trotta 2008. The validity and utility of these model selection-based approaches as applied to exoplanet atmospheres have been routinely demonstrated (Benneke and Seager 2013; Kreidberg et al. 2014a; Line et al. 2016). It is essential to use these evidence-based model selection methods to explore the hierarchy of model assumptions in order to diagnose the significance of the assumptions and/or missing model physics.

2.6.3 An Example: Cool T-Type Brown Dwarfs

While exoplanet spectra typically have low signal-to-noise spectra and spectral coverage over a limited wavelength range, brown dwarfs have excellent spectra over a broad wavelength range, and have been studied in some detail since their first discovery in 1995. An excellent review article on observations of these objects can be found in Kirkpatrick (2005). Line et al. (2014, 2015, 2017) pioneered the use of retrieval methods for these objects. T-type brown dwarfs are especially good targets because they typically lack the thick clouds in L-type brown dwarfs. This means that their photospheres span a relatively large dynamic range in pressure, meaning that information on the atmosphere can be gleaned from many levels.

As a test-case in what we might expect for exoplanets in the future, when we have *JWST*-quality (or better) data for cool atmospheres, (Line et al., 2015) studied two benchmark brown dwarfs, Gl 570D and HD 3651B, both of which are ~700 K brown dwarfs that orbit well-characterized Sunlike parent stars. Figure 2.18 shows the results of the retrievals. The top panel shows the outstanding agreement between the model-derived spectra and the medium-resolution observations. The work aimed to retrieve the atmospheric *P*–*T* profile at 15 levels in the atmosphere. The middle panel compares the derived profile to that of a best-fit radiative-convective self-consistent model atmosphere (Saumon and Marley 2008), and the agreement is striking.

Finally, the retrievals allow for the determination of chemical abundances, a first for a fit to brown dwarf spectra. The bottom panel of Fig. 2.18 shows the derived

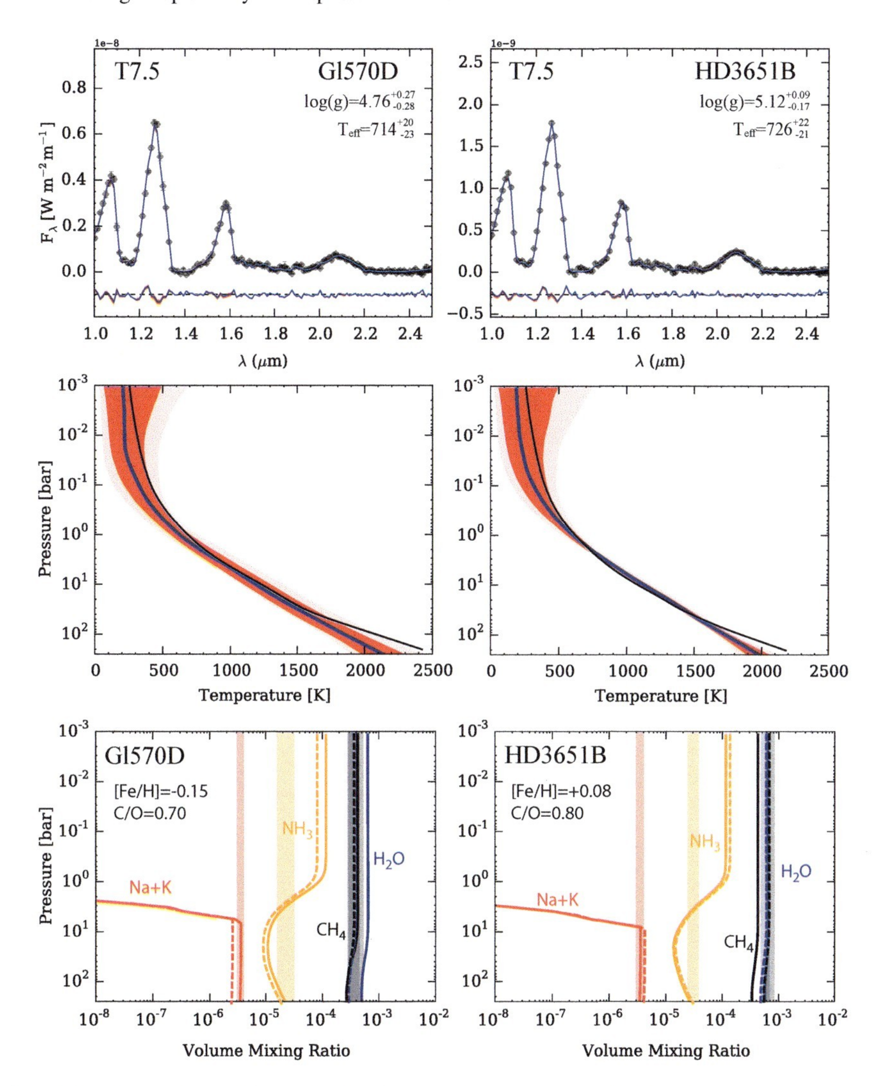

Fig. 2.18 Adapted from Line et al. (2015), retrievals on two ~700 K T-type brown dwarfs. Spectra (top row), retrieved temperature profiles (middle row), and retrieved atmospheric abundances (bottom row). *Top:* For the two objects we show the H-band calibrated SpeX spectra data as the diamonds with error bars, a summary of thousands of model spectra generated from the posterior and their residuals (median in blue, 1σ spread in red), and their spectral type and bulk properties. *Middle:* This summarizes thousands of temperature profiles drawn from the posteriors for each object (median in blue, 1σ spread in red, 2σ spread in pink). The black temperature profile shown for each object is a representative self-consistent grid model (Saumon and Marley 2008) interpolated to the quoted $\log g$ and $T_{\rm eff}$ to demonstrate that the retrieved profiles are physical and are consistent with 1D radiative-convective equilibrium. *Bottom:* Comparison of the retrieved chemical abundances (shaded boxes) of the well constrained molecules with their expected thermochemical equilibrium abundances along the median temperature profile. The solid curves are the thermochemical equilibrium abundances for solar composition while the dashed curves are the thermochemical equilibrium abundances for the specified C/O and metallicity. This shows that the retrieved abundances are thermo-chemically consistent

abundances, within the 1σ error bars, as shaded colors, with the expected chemical equilibrium abundances as solid and dashed curves. The agreement is quite good. (We note that the sharp drop in alkali metals Na and K is due to their condensation into solid cloud particles, and corresponding loss from the gas phase.) The implied abundances for carbon and oxygen, and the C/O ratio, agree well with a detailed analysis of the a high-resolution spectrum of each object's parent star. Since the brown dwarf and the Sunlike star formed in a bound orbit, within the same giant molecular cloud, they should share the same abundances. This gives confidence in the retrieval methodology.

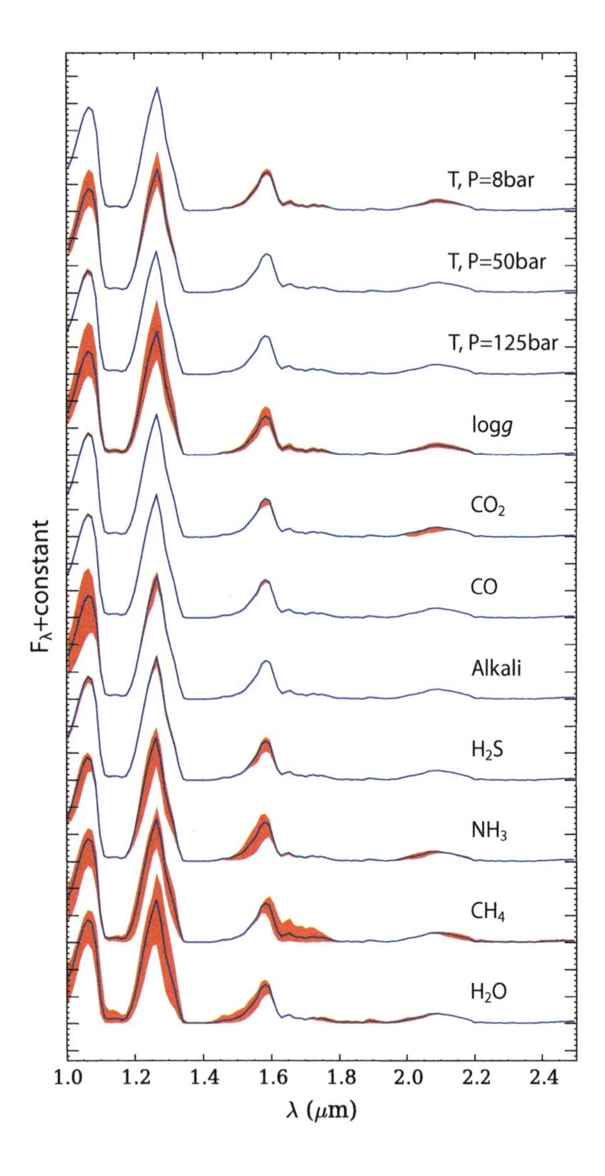

Fig. 2.19 Sensitivity of a cool brown dwarf spectrum to various parameters. This is a synthetic spectrum with T_{eff} of 700 K and a log g of 5 and with purely thermochemical equilibrium composition. The red regions represent a change in the spectrum due to a perturbation of each of the parameters. For H_2O, CH_4, NH_3, H_2S, alkali, this perturbation is $\pm$0.5 dex (where 1 dex is 1 order of magnitude) in number mixing ratio from the thermochemical abundance value. For CO the perturbation is +2 dex, CO_2, +6 dex, log$g$ $\pm$ 0.1 dex, and the temperatures are perturbed at each level by $\pm$50 K. Figure courtesy of Michael Line

Another nice feature that one can derive from retrieval models is the degree to which changes in chemical abundances, the temperature profile, or surface gravity affect the spectra at particular wavelengths. This can give one an intuitive feel for which wavelengths are most sensitive to particular aspects of the model, and why some features are essentially insensitive to the spectrum itself. This is nicely displayed in Fig. 2.19 for a generic 700 K brown dwarf model. One can first look at the three pressures, 8, 50, and 125 bars. The spectrum gives very little sensitivity to the 8-bar temperature, because the atmosphere is optically thin at most wavelengths at this pressure. However, at 50 bars, we are typically seeing down into the Y (1.0 μm), J (1.2 μm), and H (1.6 μm) band windows, so the spectrum gives us great leverage on the temperature there. By 125 bars the atmosphere is nearly opaque at all wavelengths, so we have very little leverage on the temperature at such a high pressure.

For the atmospheric surface gravity all wavelengths contribute to our understanding, as this is a constant in the atmosphere. For the molecular abundances, as one might expect, there is only "power" at wavelengths where the particular molecules are good absorbers *and* if these molecules are abundant enough to create any spectral features. At such a cool temperature (700 K), CO and CO_2 have very low mixing ratios, so we do not see them, and they have little impact on the spectrum. The pressure-broadened alkali metals, Na and K, impact the spectrum via their pressure-broadened red wing that overlaps with the Y and J bands. H_2S has little abundance so it impacts the spectrum modestly. NH_3, CH_4, and H_2O all have multiple molecular bands throughout the near-infrared, so we should expect to be able to determine their abundances relatively robustly, in agreement with the bottom panel of Fig. 2.18.

2.7 Simplified Atmospheric Dynamics

There are a number of excellent reviews of atmospheric dynamics in the exoplanet context, including Showman et al. (2008) and Heng and Showman (2015) on hot Jupiters and Showman et al. (2013) on terrestrial planets. This is a huge subject and the interested reader can find a robust literature that connects the dynamics of solar system rocky planets to solar system gas giants to exoplanets. Here we will suffice to discuss a few relevant timescales that help us understand the detected phase curves of (likely) tidally locked hot Jupiters, which are the planets whose atmospheres have been probed to understand dynamics (Fig. 2.20).

Within the realm of hot Jupiters, state-of-the-art three dimensional circulation models include a treatment of the Navier-Stokes equations (or a simplified version of them that assume hydrostatic equilibrium, called the Primitive Equations), for the fluid dynamics. This is combined with a treatment of the radiative transport in the atmosphere, which can be fully wavelength-dependent (termed "non-gray" in astronomical jargon) across a wide wavelength range from the blue optical to

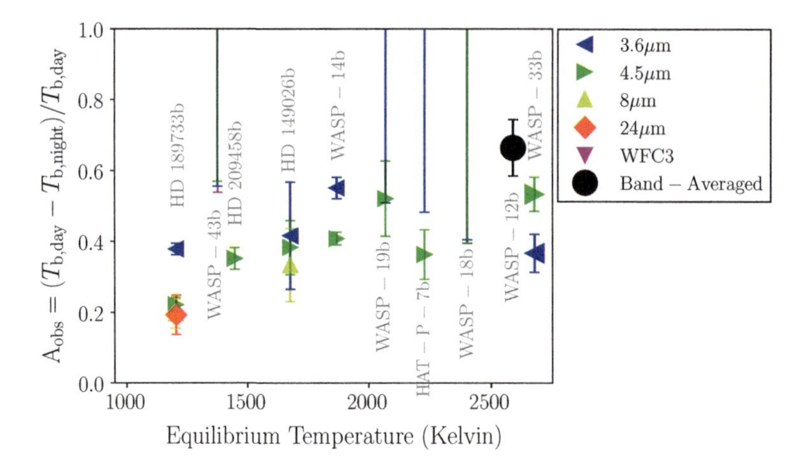

Fig. 2.20 This is a compilation of phase curve data, mostly from *Spitzer*, that shows the day/night temperature contrast for transiting giant planets. The y-axis is a measure of day/night temperature homogenization, with more homogenized planets plotting lower. There is a weak general trend that the hottest planets have the largest day/night contrast, although these inferences depend on the wavelength observed (as they probe different depths in the atmosphere). Figure courtesy of Tad Komacek

the far infrared, or can also averaged over in terms of weighted mean wavelength-independent visible ("shortwave") and corresponding thermal ("longwave") fluxes. Such "grey" simplifications speed up codes dramatically, but have significant drawbacks when comparing to wavelength-dependent thermal infrared phase curves, as can be obtained from *Spitzer* or *Hubble* (Knutson et al. 2007, 2009; Stevenson et al. 2014).

State of the art models including non-gray radiative transfer are described in Showman et al. (2009); Lewis et al. (2014); Mayne et al. (2014). However, one should be very clear that *all* levels of model sophistication are important within dynamics. Within the dynamics literature one often finds a discussion of a "hierarchy of models," from simple 1-layer models in 2D, to simplified 3D models, to 3D models with full radiative transfer. It is only through an understanding of the physical mechanisms across multiple levels of complexity that one can begin to understand the diverse physical behavior of atmospheres.

As a basic introduction, one often discusses relevant physical timescales (Showman and Guillot 2002). These include the advective timescale, the timescale over which a parcel of gas is moved within an atmosphere. This quantity is characterized by the atmospheric wind speed and a relevant planetary distance, on the order of a planetary radius. Thus, the advective time is:

$$\tau_{\text{advec}} = \frac{R_{\text{P}}}{U} \tag{2.4}$$

where R_P is the planet radius and U is the wind speed. This can be compared to a radiative timescale, the time it takes for a parcel of gas to cool off via radiation to space:

$$\tau_{rad} = \frac{P}{g}\frac{c_P}{4\sigma T^3}. \tag{2.5}$$

If $\tau_{advec} << \tau_{rad}$, we would expect temperature homogenization around the planet. This is the case for Jupiter, as τ_{rad} is quite large, owing to the cold atmospheric temperatures. If $\tau_{advec} >> \tau_{rad}$, then we should expect large temperature contrasts on the planet. All things being equal, for even hotter planets, τ_{rad} will become smaller, given the strong temperature dependence. With a relatively constant wind speed, this would imply larger temperature contrasts. This could manifest itself in large day-night temperature differences observed for the hot Jupiters, which are expected to be tidally locked. This is indeed what is observed. In addition, other physical forces, such a Lorentz drag due to the thermal ionization of alkali metals, can slow advection as well (Perna et al. 2010), also aiding large temperature differences.

2.8 Connection with Formation Models

The atmospheres of all four of the solar system giant planets are enhanced in "metals" (elements heavier than helium) compared to the Sun. Spectroscopy of these atmospheres yields carbon abundances (via methane, CH_4) that show an enrichment of ~4, 10, 80, and 80, for Jupiter, Saturn, Uranus, and Neptune, respectively. Spectra that determine other atmospheric abundances are challenging due to the very cold temperature of these atmospheres, which sequester most molecules in clouds far below the visible atmosphere. The 1995 *Galileo Entry Probe* into Jupiter measured in situ enrichments from factors of 2–5, for light elements (C, N, S, P) and the noble gases e.g., (Wong et al. 2004). Taken as a whole, these measurements suggest that giant planet atmospheres are enhanced in metals compared to parent star values.

Before the dawn of exoplanetary science, these atmospheric metal enrichments were understood within the standard "core-accretion" model of giant planet formation (e.g., Pollack et al. 1996). Within this theory, after a solid core of ice and rock attains a size of ~10 Earth masses, this core accretes massive amounts of H/He-dominated gas from the solar nebula, which can be several Earth masses to many hundreds of Earth masses. This accretion also includes solid planetesimals that are abundant within the midplane of the solar nebula disk. The accretion of solids and gas leads to an H/He envelope enriched in metals compared to parent star abundances. Since giant planet H/He envelopes are mostly or fully convective, the metal enrichment of the envelope will include the visible atmosphere.

This planet formation framework was built upon a sample size of only 4 planets. The promise of exoplanetary science is to understand metal enrichment, and its relevance to planet formation, over a vastly larger sample size. This planet *mass-metallicity relation* needs to be understood in terms of the enrichment as a function of planet mass, but also the intrinsic dispersion in the enrichment at a given mass, since it appears that exoplanet populations are quite diverse (Thorngren et al. 2016). The metal enrichment observed today drives our understanding of the accretion of gas and solids in the planet formation era.

Recently, population synthesis formation models have aimed to understand the metal enrichment of atmospheres from massive gas giants down to sub-Neptunes. An example from Fortney et al. (2013) is shown in Fig. 2.21. These models follow the accretion of gas and solids and aim to calculate the amount of solid matter accreted by the planet as well as the fraction that ablates into the atmosphere. This is quite uncertain, as the size (or size distribution) of planetesimals is unknown, and the physics of ablation in these atmospheres is still not well understood theoretically. Nonetheless, the general trend of the models agrees well with that seen in the solar system.

A recent update to the classic Pollack et al. (1996) picture is the role of condensation ("snow lines") in controlling the local composition of planet-forming disks like the solar nebula. For instance, Öberg et al. (2011) suggest that the condensation of water into solid form beyond the water snow line will drive up the C/O ratio of the local gas (since O is lost into solids) but will drive down the C/O ratio of the solids (due to the incorporation of O into solid water). This is shown graphically in Fig. 2.22 for a standard solar nebula model. The accretion of the H/He envelope of the planet, with a mix of gas and solids, could be a fingerprint of the local conditions in the disk. This has been an essential new idea in connecting atmospheric observations to planet formation.

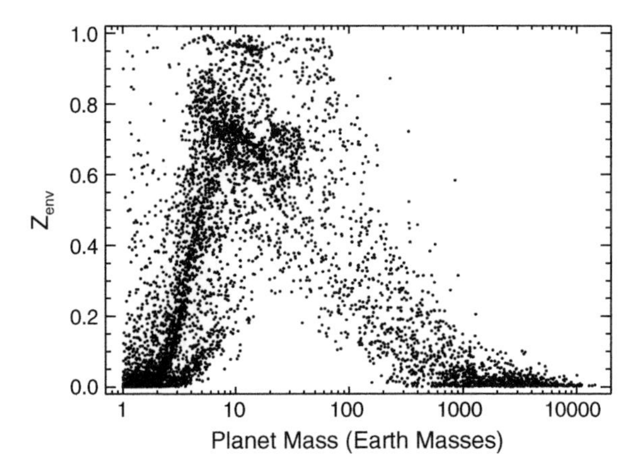

Fig. 2.21 Mass fraction of metals in the H/He envelope and visible atmosphere ($Z_{\rm env}$) as a function of planet mass for the output of the population synthesis models of Fortney et al. (2013). The dots are individually formed planets and use 100 Km planetesimals. We make a simple assumption of a uniform $Z_{\rm env}$ throughout the envelope. The turnover below 10 $M_\oplus$ is due to planetesimals driving through the atmosphere and depositing their metals directly onto the core. The general trend of the mass/atmospheric metallicity trend of the solar system system (e.g., Fortney et al. 2013; Kreidberg et al. 2014a) is reproduced by these models

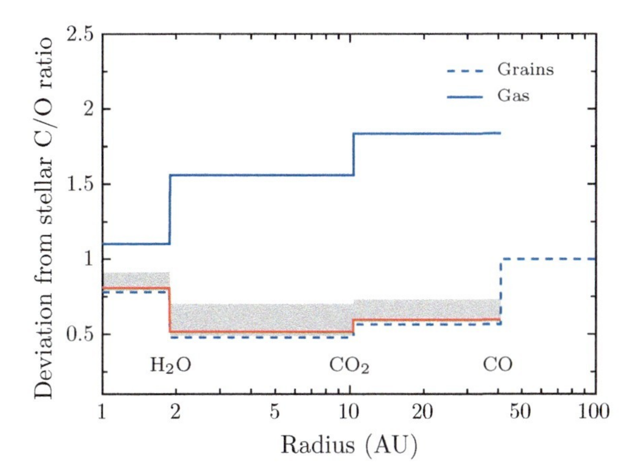

Fig. 2.22 For a typical model of the solar nebula, two values of the C/O ratio are plotted as a function of orbital separation, referenced to that of the parent star. The solid line shows the C/O ratio in the nebular gas. The dashed line shows the C/O ratios of grains, meaning the solids. Inside of 2 au, condensation of oxygen-bearing rocks removes oxygen from the gas phase, increasing the C/O of the gas. At the water snow line at 2 au, water converts to solid form, dramatically decreasing the amount of oxygen in gaseous form, therefore enhancing the C/O ratio of the gas while lowering the ratio in the solids as the amount of oxygen in solid form increases. Since giant planets accrete massive amounts of gas and solids, the final C/O ratio of the planet traces its formation location as well as its relative accretion of solids and gas. Figure courtesy of Nestor Espinoza, adapted from Espinoza et al. (2017)

This upshot is that giant planet atmospheres are likely not enriched in individual metals by uniform amounts, which is consistent with the Galileo Probe data for Jupiter. The original Öberg et al. (2011) model has since been expanded by other groups to understand the roles of additional physical processes. These processes include the accretion of solids and gas during disk-driven orbital migration, accretion via pebbles, and disk evolution/cooling over time (Madhusudhan et al. 2014, 2017; Mordasini et al. 2016; Espinoza et al. 2017). These processes all affect the location of snow lines and the composition of accretions solids. Importantly, these various theories make a wide range of predictions for the final C/O ratios of giant planets. There is broad agreement within the planet formation field that giant planets will typically not take on the same carbon and oxygen abundances as their parent star. Therefore, a derivation of the population-wide C/O ratio of the sample would be a new and unique constraint on planet formation.

2.9 Perspective

The study of exoplanetary atmospheres is still an extremely young field. There is much room for adventurous investigators in observations, theory, and modeling. We have barely scratched the surface of what there is to learn. Given the complexity of planetary atmospheres, we should not expect them to readily fall into tidy spectral classes like main sequence stars. Planets have a diversity of initial abundances, formation locations, durations of formation, and subsequent evolution in isolation or in packed planetary systems, with incident stellar fluxes of both high and low energy across a broad parent star spectral range. This rich diversity will make for a rewarding study in the near term, as *JWST* transforms the data quality that we have for studying atmospheres. In the long term, given the large phase space, the field likely will need a space telescope dedicated to obtaining high-quality spectra for a statistically large sample of atmospheres.

We should develop a hierarchy of modeling tools to confront these data sets, some simple to complex, from 1D to 3D. Within this hierarchy we should endeavor for a better understanding of molecular opacities, cloud microphysics, radiative transport, and fluid dynamics.

Acknowledgements I would like thank the organizers of the 2nd Advanced School of Exoplanetary Sciences (ASES2) for the opportunity to give lectures at the school on the beautiful Amalfi Coast. It was a fantastic experience and I benefit from great interactions with the attendees. The lectures and this chapter benefited significantly from many discussions with Mark Marley going back to 2004. Christopher Seay aided the chapter greatly by generating many of the figures. Mike Line provided figures and gave significant input on the atmospheric retrieval section. Caroline Morley helped in ways large and small. Callied Hood, Kat Feng, and Maggie Thompson provided essential comments.

References

Ackerman, A.S., Marley, M.S.: ApJ **556**, 872 (2001)
Asplund, M., Grevesse, N., Sauval, A.J., Scott, P.: ARA&A **47**, 481 (2009)
Baraffe, I., Chabrier, G., Barman, T.S., Allard, F., Hauschildt, P.H.: A&A **402**, 701 (2003)
Benneke, B., Seager, S.: ApJ **778**, 153 (2013)
Buchner, J., Georgakakis, A., Nandra, K., et al.: A&A **564**, A125 (2014)
Burrows, A., Ibgui, L., Hubeny, I.: ApJ **682**, 1277 (2008)
Burrows, A., Marley, M., Hubbard, W.B., et al.: ApJ **491**, 856 (1997)
Catling, D.C., Kasting, J.F.: Atmospheric Evolution on Inhabited and Lifeless Worlds (2017)
Chamberlain, J.W., Hunten, D.M.: Theory of planetary atmospheres. An introduction to their physics and chemistry (1987)
Espinoza, N., Fortney, J.J., Miguel, Y., Thorngren, D., Murray-Clay, R.: ApJ **838**, L9 (2017)
Feroz, F., Hobson, M.P., Bridges, M.: MNRAS **398**, 1601 (2009)
Foreman-Mackey, D., Hogg, D.W., Lang, D., Goodman, J.: PASP **125**, 306 (2013)
Fortney, J.J.: MNRAS **364**, 649 (2005)
Fortney, J.J., Ikoma, M., Nettelmann, N., Guillot, T., Marley, M.S.: ApJ **729**, 32 (2011)
Fortney, J.J., Lodders, K., Marley, M.S., Freedman, R.S.: ApJ **678**, 1419 (2008)
Fortney, J.J., Marley, M.S., Barnes, J.W.: ApJ **659**, 1661 (2007)
Fortney, J.J., Marley, M.S., Lodders, K., Saumon, D., Freedman, R.: ApJ **627**, L69 (2005)
Fortney, J.J., Mordasini, C., Nettelmann, N., et al.: ApJ **775**, 80 (2013)
Greene, T.P., Line, M.R., Montero, C., et al.: ApJ **817**, 17 (2016)
Guillot, T.: A&A **520**, A27 (2010)
Hansen, B.M.S.: ApJS **179**, 484 (2008)
Hauschildt, P.H., Allard, F., Ferguson, J., Baron, E., Alexander, D.R.: ApJ **525**, 871 (1999)

Helling, C., Tootill, D., Woitke, P., Lee, G.: A&A **603**, A123 (2017)
Helling, C., Woitke, P.: A&A **455**, 325 (2006)
Helling, C., Woitke, P., Thi, W.-F.: A&A **485**, 547 (2008)
Heng, K.: Exoplanetary Atmospheres: Theoretical Concepts and Foundations (2017)
Heng, K., Marley, M. (2017). arXiv:1706.03188
Heng, K., Showman, A.P.: Annu. Rev. Earth Planet. Sci. **43**, 509 (2015)
Hubeny, I.: MNRAS **469**, 841 (2017)
Kirkpatrick, J.D.: ARA&A **43**, 195 (2005)
Knutson, H.A., Charbonneau, D., Allen, L.E., et al.: Nature **447**, 183 (2007)
Knutson, H.A., Charbonneau, D., Cowan, N.B., et al.: ApJ **690**, 822 (2009)
Kreidberg, L., Bean, J.L., Désert, J.-M., et al.: ApJ **793**, L27 (2014a)
Kreidberg, L., Bean, J.L., Désert, J.-M., et al.: Nature **505**, 69 (2014b)
Kreidberg, L., Line, M.R., Bean, J.L., et al.: ApJ **814**, 66 (2015)
Lewis, N.K., Showman, A.P., Fortney, J.J., Knutson, H.A., Marley, M.S.: ApJ **795**, 150 (2014)
Line, M.R., Fortney, J.J., Marley, M.S., Sorahana, S.: ApJ **793**, 33 (2014)
Line, M.R., Teske, J., Burningham, B., Fortney, J.J., Marley, M.S.: ApJ **807**, 183 (2015)
Line, M.R., Wolf, A.S., Zhang, X., et al.: ApJ **775**, 137 (2013)
Line, M.R., Stevenson, K.B., Bean, J., et al.: AJ **152**, 203 (2016)
Line, M.R., Marley, M.S., Liu, M.C., et al.: ApJ **848**, 83 (2017)
Lodders, K., Fegley, B.: Icarus **155**, 393 (2002)
Lopez, E.D., Fortney, J.J.: ApJ **792**, 1 (2014)
Lupu, R.E., Marley, M.S., Lewis, N., et al.: AJ **152**, 217 (2016)
Lupu, R.E., Zahnle, K., Marley, M.S., et al.: ApJ **784**, 27 (2014)
Madhusudhan, N., Amin, M.A., Kennedy, G.M.: ApJ **794**, L12 (2014)
Madhusudhan, N., Bitsch, B., Johansen, A., Eriksson, L.: MNRAS **469**, 4102 (2017)
Madhusudhan, N., Mousis, O., Johnson, T.V., Lunine, J.I.: ApJ **743**, 191 (2011a)
Madhusudhan, N., Harrington, J., Stevenson, K.B., et al.: Nature **469**, 64 (2011b)
Marley, M.S., Fortney, J.J., Hubickyj, O., Bodenheimer, P., Lissauer, J.J.: ApJ **655**, 541 (2007)
Marley, M.S., Gelino, C., Stephens, D., Lunine, J.I., Freedman, R.: ApJ **513**, 879 (1999)
Marley, M.S., Robinson, T.D.: ARA&A **53**, 279 (2015)
Marley, M.S., Saumon, D., Cushing, M., et al.: ApJ **754**, 135 (2012)
Marley, M.S., Saumon, D., Guillot, T., et al.: Science **272**, 1919 (1996)
Mayne, N.J., Baraffe, I., Acreman, D.M., et al.: A&A **561**, A1 (2014)
McKay, C.P., Pollack, J.B., Courtin, R.: Icarus **80**, 23 (1989)
Mollière, P., van Boekel, R., Dullemond, C., Henning, T., Mordasini, C.: ApJ **813**, 47 (2015)
Mordasini, C., van Boekel, R., Mollière, P., Henning, T., Benneke, B.: ApJ **832**, 41 (2016)
Morley, C.V., Fortney, J.J., Marley, M.S., et al.: ApJ **756**, 172 (2012)
Morley, C.V., Fortney, J.J., Marley, M.S., et al.: ApJ **756**, 172 (2013)
Morley, C.V., Fortney, J.J., Marley, M.S., et al.: ApJ **815**, 110 (2015)
Morley, C.V., Knutson, H., Line, M., et al.: AJ **153**, 86 (2017a)
Morley, C.V., Kreidberg, L., Rustamkulov, Z., Robinson, T., Fortney, J.J.: ApJ **850**, 121 (2017b)
Öberg, K.I., Murray-Clay, R., Bergin, E.A.: ApJ **743**, L16 (2011)
Parmentier, V., Guillot, T.: A&A **562**, A133 (2014)
Perna, R., Menou, K., Rauscher, E.: ApJ **724**, 313 (2010)
Pierrehumbert, R.T.: Principles of Planetary Climate (2010)
Pollack, J.B., Hubickyj, O., Bodenheimer, P., et al.: Icarus **124**, 62 (1996)
Robinson, T.D., Catling, D.C.: ApJ **757**, 104 (2012)
Saumon, D., Marley, M.S.: ApJ **689**, 1327 (2008)
Seager, S.: Exoplanet Atmospheres: Physical Processes (2010)
Seager, S., Deming, D.: ARA&A **48**, 631 (2010)
Showman, A.P., Fortney, J.J., Lian, Y., et al.: ApJ **699**, 564 (2009)
Showman, A.P., Guillot, T.: A&A **385**, 166 (2002)

Showman, A.P., Menou, K., Cho, J.Y.-K.: in Fischer, D., Rasio, F.A., Thorsett, S.E., Wolszczan, A. (eds.) Astronomical Society of the Pacific Conference Series. Astronomical Society of the Pacific Conference Series, vol. 398, p. 419–+ (2008)
Showman, A.P., Wordsworth, R.D., Merlis, T.M., Kaspi, Y.: in Mackwell, S.J., Simon-Miller, A.A., Harder, J.W., Bullock, M.A. (eds.) Atmospheric Circulation of Terrestrial Exoplanets, pp. 277–326 (2013)
Sing, D.K., Fortney, J.J., Nikolov, N., et al.: Nature **529**, 59 (2016)
Stevenson, K.B., Désert, J.-M., Line, M.R., et al.: Science **346**, 838 (2014)
Sudarsky, D., Burrows, A., Pinto, P.: ApJ **538**, 885 (2000)
Thorngren, D.P., Fortney, J.J. (2017). arXiv:1709.04539
Thorngren, D.P., Fortney, J.J., Murray-Clay, R.A., Lopez, E.D.: ApJ **831**, 64 (2016)
Toon, O.B., McKay, C.P., Ackerman, T.P., Santhanam, K.: Journal of Geophysical Research **94**, 16287 (1989)
Trotta, R.: Contemporary Physics **49**, 71 (2008)
Wong, M.H., Mahaffy, P.R., Atreya, S.K., Niemann, H.B., Owen, T.C.: Icarus **171**, 153 (2004)

Part III
Molecular Spectroscopy

Chapter 3
Molecular Spectroscopy for Exoplanets

Jonathan Tennyson

Abstract One can determine orbital radius, planetary size and planetary mass, and hence average density from techniques used to detect exoplanets. The only real means of expanding on this information is by spectroscopic study of atmospheres. To do this requires a knowledge of the spectroscopy of the molecules, present or likely to be present, in their atmosphere which in turn demands access to the extensive laboratory data that characterises these species. Given that the exoplanets whose spectra can be observed are largely hot compared to our planet, this puts particular demands on spectroscopic data needed both to interpret any observations and to perform radiative transport models of planets of interest. This chapter outlines the basic spectroscopy of atoms and molecules with a particular emphasis on the molecules that likely to form in exoplanetary atmospheres. The importance of treating the temperature dependence of the spectrum and the huge growth in the number of lines which play a role at higher temperatures, such as those deduced for hot Jupiter exoplanets, is emphasized. Sources of data for use in studies of exoplanets are discussed in detail and illustrative examples given.

3.1 General Introduction

The last two decades have taught us that exoplanets are common; indeed it seems that nearly every star supports a planetary system. Logically the next step is to understand more about these newly-discovered bodies. One can determine orbital radius, planetary size and planetary mass, and hence average density from techniques used to make the original detections. The only real means of expanding on this information is by spectroscopic probes of the atmosphere. To do this requires a knowledge of the species, largely molecules, present or likely to be present in the atmosphere. It also requires access to the laboratory data that characterises these species. Given that the exoplanets whose spectra can be observed are largely hot

---

J. Tennyson (✉)
Department of Physics and Astronomy, University College London,
London WC1E 6BT, UK
e-mail: j.tennyson@ucl.ac.uk

V. Bozza et al. (eds.), *Astrophysics of Exoplanetary Atmospheres*, Astrophysics and Space Science Library 450, https://doi.org/10.1007/978-3-319-89701-1_3

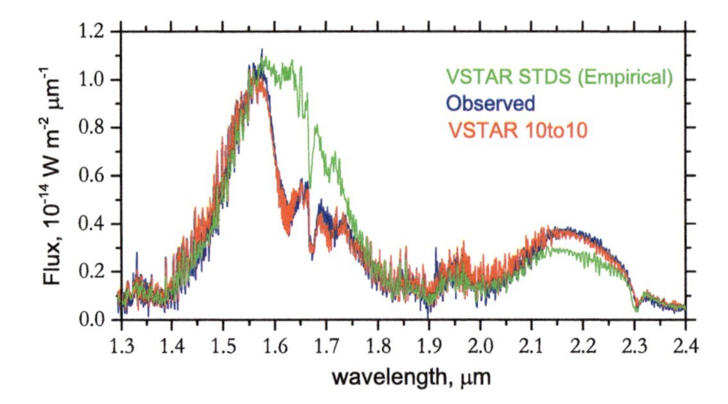

Fig. 3.1 Spectra of the T 4.5 dwarf 2M 0559-14. The observed spectrum (blue line) was taken with the SpeX instrument on the 3-m NASA Infrared Telescope Facility (IRTF). Models were calculated with VSTAR for $T = 1500$ K using **a** the empirical STDS line list of Wenger and Champion (1998) and **b** the computed 10to10 line list of Yurchenko and Tennyson (2014). The figure is adapted from the work of Yurchenko et al. (2014)

compared to our planet, this puts particular demands on spectroscopic data needed both to interpret any observations and to perform radiative transport models of planets of interest.

Figure 3.1 gives an illustrative example of the importance of using comprehensive sets of spectroscopic data. The example is actually for a methane-rich T-dwarf which provides a better text case since the current status of exoplanet spectroscopy is not good enough to provide a stringent test. The empirical data (STDS) contains a few 100,000 lines while a computed line list contains almost 10^{10} lines. It was found that good results could only be obtained with the inclusion of about 3 billion lines. The need for comprehensive datasets is clear.

This chapter concentrates heavily, but not exclusively, on the spectroscopy of atoms and molecules likely to be found in the atmospheres of exoplanets. A more general introduction to astronomical spectroscopy can be found in my book (Tennyson 2011) of that title. Molecular spectra are governed by the rules of quantum mechanics; my book starts from a basic knowledge of the quantum mechanics of the hydrogen atom to build the rules that govern astronomical spectra and uses observed spectra to illustrate the results. Some knowledge of basic quantum mechanics will be assumed in this chapter. I will survey the underlying nature of molecular spectra giving an idea of the information required to understand them. This survey does little more than give a flavour of the issues involved; for the reader wishing get a deeper or more specialised understanding there are a number of comprehensive text books available of which the book "Spectra of Atoms and Molecules" by Bernath (2015) is an excellent example.

3.1.1 The Basics

Spectroscopy of atoms and molecules probes the energy level structure of these species by absorption or emission of quanta of light, known as photons. These photons possess precisely the right energy to bridge the gap between two energy levels; this is depicted schematically in Fig. 3.2. Because the energy level structure of each atom, ion or molecule is different, the wavelengths of the light absorbed or emitted are unique to that species allowing its identification even over vast distances.

Absorption is the commonest technique used in laboratory high resolution spectroscopy experiments; it requires an appropriate source of light at the wavelengths of interest. This can be hard to arrange in astrophysics but for exoplanets the host star provides a natural light source allowing, for example, absorption spectra to be recorded when a transiting planet moves in front of its host star. Conversely, emission spectra generate their own light by the emitting energy levels becoming overpopulated by thermal or some other means of excitation. The so-called secondary transit probes emissions from a planet's atmosphere.

There is a vast wealth of laboratory spectroscopic data compiled over decades from detailed, high-quality laboratory experiments and, increasingly, from theoretical calculations. Sources of these data are discussed at the end of this chapter. However, exoplanet spectroscopy has raised new challenges for the provision of laboratory data. This is not so much because of the unusual nature of the species: most of the molecules detected, or searched for can be considered to be the usual suspects whose spectra are well-studied on Earth. The issue is to with the environment: first

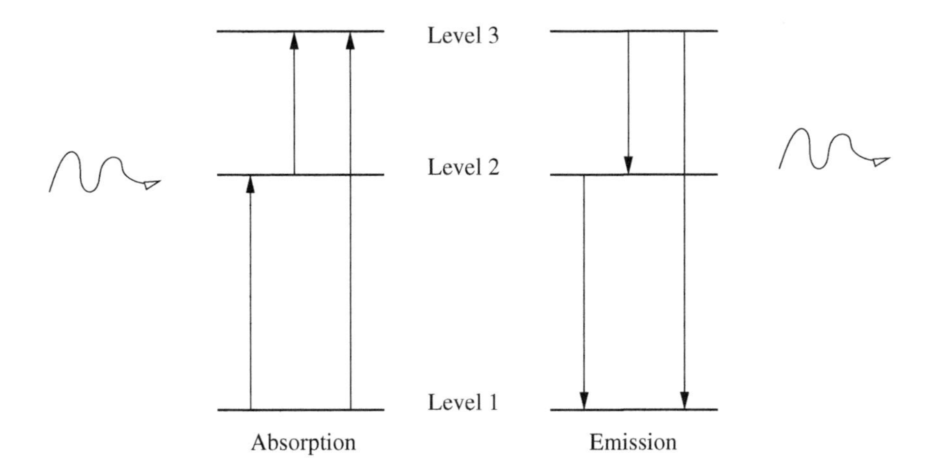

Fig. 3.2 Schematic three-level system showing absorption of an incoming photon (left) or emission of an outgoing photon (right) between the levels. The diagram assumes that transitions between all three levels are allowed, which is often not the case. Note that emission and absorption occur at the same photon wavelength/frequency and that the energy of a photon linking levels 1 and 3 is precisely the sum of the energies of the two photons that link levels 1 and 2, and levels 2 and 3

the exoplanets for which there is any immediate hope of obtaining good spectra are all hot which, as discussed below, can hugely increase the amount of laboratory data needed to analyse or interpret spectra; secondly, at least for gas giant planets, the atmospheres are assumed to be largely composed of molecular hydrogen and atomic helium meaning that collisions effects (pressure broadening) on the spectra of molecules contained therein differs significantly from effects observed in our own atmosphere.

3.1.2 Spectroscopic Regions and Units

Wavelength (λ) and frequency (ν) are related by the speed of light (c) by $\lambda = \frac{c}{\nu}$. Astronomers generally use wavelengths, which have units of length (μm, Å, nm, etc.), in the infrared and visible range as telescopes naturally work in wavelengths. However at the extremes of the spectrum they generally use frequency or energy units, which are closely related. In short wavelength regions, extreme ultraviolet (EUV), X-ray and γ-ray, energy units (usually electron Volts or eV) are generally employed. At long wavelengths such as the radio region which is used to probe the interstellar medium, spectra are usually recorded using frequencies (Hz) although in practice they are often presented as the Doppler shift (in km s^{-1}) from a known laboratory feature. Although this makes sense from the point of view of interpreting motions of clouds etc., it is somewhat bizarre from a spectroscopic viewpoint.

The physics of spectra are most easily understood using energies because this gives a simple relationship between the energy levels and the properties of the transitions. Thus laboratory spectra often use Hz at long wavelengths and eV at very short wavelengths. In the intermediate region, which is the one of most importance for exoplanets, the rather awkward wavenumber units of cm^{-1} are uniformly used. When converting it is worth remembering that a transition at a wavelength of 1 µm has a wavenumber of 10 000 cm^{-1} and they are, of course, inversely related. Figure 3.3 gives an overview of various different units and allows a very approximate conversion between them. Finally it should be noted that astronomers often given energies expressed in units of absolute temperature (K). In this context it is worth noting that 1 K $\approx$ 0.69 cm^{-1}. However one should be careful here because a temperature represents a distribution of energies whereas energy levels and transitions occur at precise values.

Figure 3.3 also shows the various spectroscopic regions of the electromagnetic spectrum. As discussed below the physics of various molecular transitions determines the region in which they can be observed. Conversely the observing strategy is more determined by the spectroscopic behaviour of the Earth's own atmosphere and, of course, there can be issues trying to observe the atmosphere of one planet from inside the atmosphere of another one.

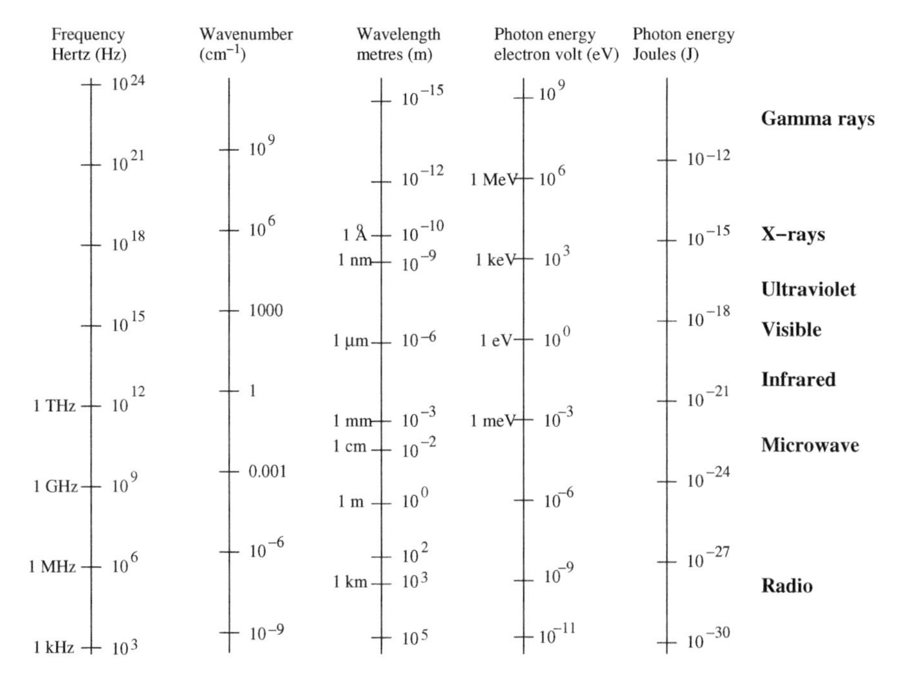

Fig. 3.3 The electromagnetic spectrum

3.1.3 What Does One Learn from an Astronomical Line Spectrum?

The analyses of spectra have provided a wealth of information on the Universe about us. Careful study of spectroscopic data, particularly high-resolution data, can be rich in information about both the chemical and physical characteristics of the object under observation. In particular on can use spectra to determine:

- **Which species produces the line(s)**? This provides information on the chemical composition of the object under observation.
- **Which transition is it**? The comparison of occupancy of ground and excited states gives information about the local physical conditions of the object such as temperature.
- **How strong is the line**? Knowledge of thermal population and transition intensities can be used to reveal the abundance or total column of the species under observation.
- **What is the Doppler shift**? Doppler shifts give information on the motion of the object or region of the object. At high resolution, these data can provide a wealth of information about the local environment of the species.

- **What is the line profile**? Line profiles provide information of the local physical conditions; in principal both the (translational) temperature and the local pressure can be determined from the Doppler and pressure broadening, respectively. Line profiles also provide information on the optical depth.
- **What is the splitting pattern**? Local magnetic fields can be observed via the splitting of spectral lines by the field.

While temperature is one of the most important parameters retrieved from spectral observations, it must be borne in mind that most of the Universe is not in local thermodynamic equilibrium (LTE). Even if observations are well-modeled by a single temperature, this temperature may just represent the effective temperature of a single type of motion such as translation or rotation. Molecules are complicated systems and it is possible to use spectroscopy to determine more than one temperature, see Dello Russo et al. (2005) for example. These different temperatures in turn provide important information about thermalisation timescales.

3.2 Atomic Spectra

Hot objects such as stars with solar temperature ($\approx$5600 K) and above are primarily composed of atomic species at various stages of ionisation. The spectra of these objects are important in their own right as well providing the back drop for exoplanet observations.

In general atomic spectra involve the movement of electrons; one astronomically important exception being the H atom 21 cm line, discussed below, which involves a nuclear spin flip. These electronic transitions are generally strongest at ultraviolet and visible wavelengths. The transitions are observed as discrete lines or groups of lines unlike the bands observed in hot molecules. Nearly all stellar spectra contain prominent atomic features.

All spectroscopic transitions obey selection rules which govern the energy levels which can be reached starting from a given level. As atoms are high symmetry objects they are characterised by a series of quantum numbers which define their states. A discussion of the physics behind these quantum numbers and the rather arcane notation used to represent them can be found in standard textbooks such as Woodgate (1983) and Tennyson (2011).

Table 3.1 summarises the selection rules which give atomic spectra. Electric dipole transitions are generally very much stronger than electric quadrupole or magnetic dipole transitions and, as result, only these transitions are important in the spectra of stars and exoplanets. All dipole allowed transitions must obey the first three selection rules listed but the strongest features, which are the ones usually observed generally obey all the rules listed.

Dipoles usually given in the cgs unit Debye (D). At atomic level, the separation of a unit charge by 1 Bohr (a typical atomic or molecular scale) gives a dipole of 2.54 D.

Table 3.1 Selection rules for atomic spectra. Rules 1, 2 and 3 must always be obeyed. For electric dipole transitions, intercombination lines violate rule 4 and forbidden lines violate rule 5 and/or 6. Electric quadrupole and magnetic dipole transitions are also described as forbidden

	Electric dipole	Electric quadrupole	Magnetic dipole
1.	$\Delta J = 0, \pm 1$	$\Delta J = 0, \pm 1, \pm 2$	$\Delta J = 0, \pm 1$
	Not $J = 0 - 0$	Not $J = 0 - 0, \frac{1}{2} - \frac{1}{2}, 0 - 1$	Not $J = 0 - 0$
2.	$\Delta M_J = 0, \pm 1$	$\Delta M_J = 0, \pm 1, \pm 2$	$\Delta M_J = 0, \pm 1$
3.	Parity changes	Parity unchanged	Parity unchanged
4.	$\Delta S = 0$	$\Delta S = 0$	$\Delta S = 0$
5.	One electron jumps	One or no electron jumps	No electron jumps
	Δn any	Δn any	$\Delta n = 0$
	$\Delta l = \pm 1$	$\Delta l = 0, \pm 2$	$\Delta l = 0$
6.	$\Delta L = 0, \pm 1$	$\Delta L = 0, \pm 1, \pm 2$	$\Delta L = 0$
	Not $L = 0 - 0$	Not $L = 0 - 0, 0 - 1$	

3.2.1 Atomic Hydrogen

As the dominant species in the Universe, the spectrum of atomic hydrogen is of particular importance. H atom spectra obey the selection rules listed in Table 3.1 but, because the H atom only has a single electron this can be significantly simplified. For example, all H-atom states are spin doublets meaning $S = 2$ so the spin selection rule, $\Delta S = 0$ is always satisfied for all transitions.

H atom spectra are divided up into well-known series classified according to the lower energy level involved. These series, which played a crucial role in the early development of quantum mechanics, are listed in Table 3.2. H atom spectra were observed in exoplanets by Vidal-Madjar et al. (2003) via the so-called Lyman α transition which is the transition from the $n = 1$ ground state to the $n = 2$ first excited state. Observation of these lines need to be made in the ultraviolet. Easier to observe from the ground are Balmer series lines; however these transitions start from the excited $n = 2$ state of the H atom which means that the hydrogen must be excited by some mechanism. So far, as discussed by Cauley et al. (2017), detection of Balmer lines has been patchy in exoplanet atmospheres.

Even though every H atom has an electron spin angular momentum of a half, this angular momentum still plays an important role in H-atom spectra. The is because the spin angular moment, $s = \frac{1}{2}$ can couple with the orbital angular momentum, ℓ. For states with $\ell > 0$ (non s states), the resulting total angular momentum, j, can take two, half-integer values given by $\ell + \frac{1}{2}$ and $\ell - \frac{1}{2}$. The splitting of the spectral lines that results from the splitting of the energy levels is known as fine structure. The fine structure effect in atomic hydrogen is summarised in Table 3.3.

There is one more potential source of angular moment in an atom and this is the nucleus. If the nucleus has a non-zero spin then this can also couple with the total electronic angular momentum, j to give the final angular momentum, f.

Table 3.2 Spectral series of the H atom. Each series comprises the transitions $n_2 - n_1$, where $n_1 < n_2 < \infty$

n_1	Name	Symbol	Spectral region	Range/cm^{-1}		
				$n_2 = n_1 + 1$		$n_2 = \infty$
1	Lyman	Ly	Ultraviolet	82257	–	109677
2	Balmer	H	Visible	15237	–	27427
3	Paschen	P	Infrared	5532	–	12186
4	Brackett	Br	Infrared	2468	–	6855
5	Pfund	Pf	Infrared	1340	–	4387
6	Humphreys	Hu	Infrared	808	–	3047

Table 3.3 Fine structure effects in the hydrogen atom: splitting of the nl orbitals due to fine structure effect for $l = 0,1,2,3$. The resulting levels are labelled using H atom notation, and the more general spectroscopic notation of terms and levels (see Sect. 3.4.8)

Configuration	l	s	j	H atom	Term	Level
ns	0	$\frac{1}{2}$	$\frac{1}{2}$	$n\mathrm{s}_{\frac{1}{2}}$	$n\,^2\mathrm{S}$	$n\,^2\mathrm{S}_{\frac{1}{2}}$
np	1	$\frac{1}{2}$	$\frac{1}{2}, \frac{3}{2}$	$n\mathrm{p}_{\frac{1}{2}}, n\mathrm{p}_{\frac{3}{2}}$	$n\,^2\mathrm{P^o}$	$n\,^2\mathrm{P^o_{\frac{1}{2}}}, n\,^2\mathrm{P^o_{\frac{3}{2}}}$
nd	2	$\frac{1}{2}$	$\frac{3}{2}, \frac{5}{2}$	$n\mathrm{d}_{\frac{3}{2}}, n\mathrm{d}_{\frac{5}{2}}$	$n\,^2\mathrm{D}$	$n\,^2\mathrm{D}_{\frac{3}{2}}, n\,^2\mathrm{D}_{\frac{5}{2}}$
nf	3	$\frac{1}{2}$	$\frac{5}{2}, \frac{7}{2}$	$n\mathrm{f}_{\frac{5}{2}}, n\mathrm{f}_{\frac{7}{2}}$	$n\,^2\mathrm{F^o}$	$n\,^2\mathrm{F^o_{\frac{5}{2}}}, n\,^2\mathrm{F^o_{\frac{7}{2}}}$

This splitting, which is usually very small, is called hyperfine structure. For the H atom the nuclear spin, i, is a half. The H atom ground state, the 1 s state, has $j = \frac{1}{2}$ and coupling with i gives allowed f of 0 and 1. This splitting is very small but it is still possible to observe the very weak $f = 0 - 1$ line which lies in the radio region with a wavelength of $\lambda = 21$ cm. The 21 cm line is arguable the most important single line in astronomy. It can be used to map H distributions and velocities throughout the Universe. Its weakness means that the line is rarely saturated which allows H atom columns to be recovered with confidence.

3.2.2 *Helium Spectra*

Helium is of course the second most abundant species in the Universe. However observing it is much more difficult than for H. In a large part this is due to its atomic structure. He is a closed shell species with two electrons in the lowest orbital ($1s^2$ in standard atomic physics notation). The leading, 'resonance', line is the $1s2p - 1s^2$ transition that occurs at about 20 eV (more precisely at $\lambda = 584$ Å). This lies well into the ultraviolet requiring both a high energy light source to drive the transition and making its observation from the ground impossible. Furthermore, the line is obscured by interstellar H atoms which absorb at 584 Å. It will therefore be very hard to perform a direct, spectroscopic detection of He in the atmosphere of an exoplanet.

3.2.3 Complex Atoms

The spectra of many-electron atoms are more complicated than of hydrogen and the full atomic selection rules, as given in Table 3.1, come into play. For stellar and exoplanet spectra of these species, magnetic dipole and electric quadrupole transitions are so weak that they can safely be ignored. Electric dipole transitions can be divided into three classes. Strong, fully-allowed transitions obey all the rules specified. However, strictly dipole transitions only have to obey the first three selection rules concerning changes in ΔJ, ΔM_J and parity. The final rule is often known as the Laporte rule after its discoverer.

Transitions which obey all six selection rules are known as "allowed". Lines which satisfy all the rules except spin selection rule 4 ($\Delta S = 0$) are called intercombination lines and lines which violate selection rule 5 and/or 6 are known as "forbidden" lines. This last nomenclature is somewhat confusing as forbidden lines do occur and indeed give strong emissions in interstellar environments. However, they are generally too weak to concern us here. Similarly, intercombination lines, while generally stronger than forbidden lines are also too weak to be significant in planetary or stellar atmospheres.

Besides atomic hydrogen discussed above the main atoms detected in exoplanet atmospheres are the alkali metal atoms Na and K. For example, Pont et al. (2013) give a robust detection on exoplanet on HD189733b of Na and K lines at 0.589 μm and 0.769 μm, respectively. Alkali metals have a single "valence" electron outside a closed shell core and it is the movements this electron, alone, which gives alkali atoms their characteristic spectra.

Table 3.4 summarises the various spectral lines grouped into series. The names of these series are the origin of a nomenclature which denotes atomic orbital with 0, 1, 2 and 3 units of angular momentum s, p, d and f, respectively. The spectrum of K has a similar structure just shifted to sightly longer wavelengths. Conversely the atomic ion Mg^+ (or Mg II in standard atomic physics notation) is isoelectronic (has the same number of electrons) as Na so also displays a spectrum with very similar structure but shifted to significantly shorter wavelengths. Transitions from these species are also prominent in stellar atmospheres.

An important feature shown in Table 3.4 is the splitting of lines due to spin-orbit effects. These doublets and triplets are usually resolved in astronomical objects and can be used to determine effects such as optical thickness as the relative strengths of different transitions within a multiplet is determined by the underlying atomic physics when the lines are observed optically thin.

Table 3.4 Spectral series of sodium

Series name	Transitions	n values	Multiplicity
Sharp	$n\,^2\mathrm{S}_{\frac{1}{2}} \rightarrow 3\,^2\mathrm{P}^{\mathrm{o}}_{\frac{3}{2},\frac{1}{2}}$	$n = 4, 5, 6, \ldots$	Doublets
Principal	$n\,^2\mathrm{P}^{\mathrm{o}}_{\frac{3}{2},\frac{1}{2}} \rightarrow 3\,^2\mathrm{S}_{\frac{1}{2}}$	$n = 3, 4, 5, \ldots$	Doublets
Diffuse	$n\,^2\mathrm{D}_{\frac{5}{2},\frac{3}{2}} \rightarrow 3\,^2\mathrm{P}^{\mathrm{o}}_{\frac{3}{2},\frac{1}{2}}$	$n = 3, 4, 5, \ldots$	Triplets
Fundamental	$n\,^2\mathrm{F}^{\mathrm{o}}_{\frac{7}{2},\frac{5}{2}} \rightarrow 3\,^2\mathrm{D}_{\frac{5}{2},\frac{3}{2}}$	$n = 4, 5, 6, \ldots$	Triplets

3.2.4 Atoms (and Molecules) in Magnetic Fields

The presence of a magnetic field can cause splitting in observed spectra of atoms and molecules. These spectra, in which splittings are not always fully resolved, therefore provide the means to determine the strength of the local magnetic field. While all atomic and molecular spectra are affected by a magnetic field to some extent, the strongest effects are observed in the spectra of open shell atoms such as the spectra of Na and K discussed above.

For a so-called weak magnetic field, B, the splitting of the spectral lines is known as the Zeeman effect and takes the fairly simple form:

$$\Delta E = M_J \mu_B g_J B, \tag{3.1}$$

where μ_B is the Bohr Magneton, the basic quantum of magnetic effects, which is defined by

$$\mu_B = \frac{e\hbar}{2m_e} \tag{3.2}$$

where e and m_e are the charge and mass of the electron. and $\hbar$ is Planck's constant divided by 2π; this means $\mu_B = \frac{1}{2}$ in atomic units. J is the total angular momentum of the system and M_J is projection of J along the direction of the magnetic field; M_J takes $(2J + 1)$ values for $-J \leq M_J \leq J$ in steps of 1. Finally g_J is the so-called Landé g-factor which captures the strength of the interaction of the levels of the particular species with the magnetic field. For closed shell molecules such as H_2O, CH_4 etc., g_J is very small; it is more significant for open shell diatomics such as TiO, VO, C_2. As mentioned above, the largest g-factors are generally for open shell atoms. Introducing a magnetic field results in splitting of the levels characterised by M_J. Transitions between these levels are governed by the rule $\Delta M_J = 0, \pm 1$, see Table 3.1. Atomic spectra have widely been used to characterise the magnetic fields present in stars. People interested in a detailed exposition of this topic should look at the work of Berdyugina and Solanki (2002). Similar studies should become possible for exoplanets as spectral resolution improves.

Strong magnetic fields have a more radical effect on the level structure of atoms and molecules. This is governed by the so-called Paschen-Back effect. A discussion of the role and study of this effect in astrophysics is given by Berdyugina et al. (2005).

3.3 Molecular Motions

The spectra of molecules are more complicated and richer than those of atoms. Understanding them involves probing deeper into underlying physics as the changes induced by absorbing or emitting a photon now involve movement of the nuclei as well as, in some cases, changes in the electronic structure. Before considering the different types of molecular spectra, it is necessary to consider the different degrees of freedom available to the nuclei in a given molecule.

Consider a molecule containing N atoms. In simple Cartesian space each atom has 3 degrees of freedom so the molecule as a whole has $3N$ degrees of freedom. The bonding in the molecule puts constraints on the motions of the individual atoms and it is better to consider these degrees of freedom as a collective property of the whole molecule rather than of individual atoms. Doing this gives the following.

The translation of the whole molecule through space can be undertaken in 3 dimensions so there are 3 translational degrees of freedom. These yield essentially a continuous set of translational levels which are not of interest spectroscopically. Fortunately it is reasonably straightforward to re-write the molecular Hamiltonian to formally separate this translation motion and from here on I will assume that this has been done.

In addition to translation, the whole molecule can rotate freely in space. If the molecule is linear it can do this only the two directions perpendicular to linear structure; otherwise it will rotate in all 3 dimensions. Unlike the translation motion, the rotational motion is quantized and can be probed spectroscopically. This is discussed below.

Having identified 6 (5 for a linear molecule) degrees of freedom associated with the collective motion of the entire system, there remain $3N - 6$ ($3N - 5$ for a linear molecule) degrees of freedom which are associated with internal motions of the system. These motions are called vibrations and involve the atoms vibrating against each other. These motions are generally represented using $3N - 6$ ($3N - 5$) internal coordinates. For diatomic molecules this reduces simply to a single internal coordinate which is the one used to represent the well-known potential energy curves describing the interaction between the two atoms in the molecules. Examples of these are given below.

As for atoms, spectra can also arise due to changes (jumps) in the electronic structure of a molecule. In general these electronic changes are similar to those found in atoms but with complications due to the lower symmetry and the fact that the nuclear motion also participates in electronic transitions.

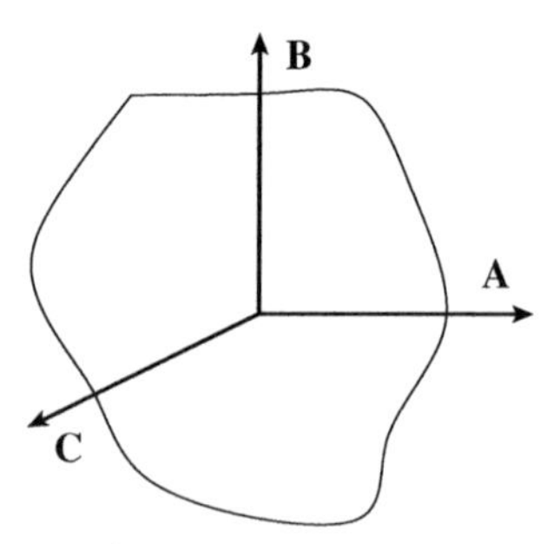

Fig. 3.4 Schematic diagram illustrating the Cartesian principal axes of rotation for an arbitrary solid body. By convention the axis with the smallest moment of inertia is labelled A and the one with the largest moment inertia is labelled. C

So for molecules there are three distinct quantized motions that can give rise to spectra: rotational motion, vibrational motion and electronic motion. Each of these possible spectra, which are discussed each in turn below, lie in a different, characteristic spectral region.

With the exception of the spectra associated with the abundant molecular hydrogen, H_2, it is only necessary to consider molecular transitions governed by electric dipole selection rules. However, molecules have an extra source of dipole moment in form of the configuration of the various nuclei in the molecule which can give rise to both permanent and instantaneous dipoles which then facilitate transitions.

3.4 Rotational Spectra

The simplest spectrum to arise from molecules is the rotational spectrum. Studies of these spectra have provided a vast amount of information of the interstellar medium and star-forming regions.

The simplest way to understand rotational spectra is to start by considering the molecule as a rigid rotor. A rigid system rotates about its so-called principal axes, see Fig. 3.4. These axes are obtained, at least in principle, by diagonalising its moment of inertia tensor. This may sound complicated but for many small molecules, such as water, the orientation of the principle axes is intuitive as it is completely determined by symmetry. Principal axes are used to represent both the classical and quantum mechanical rotational motion of solid bodies.

If the body is rigid then quantum mechanically the rotational energies levels of the system can be characterized by the rotational constant which is directly related to the component of the moment of inertia, I_X, associated with the given principal axis:

$$B = \frac{\hbar^2}{I_B}. \tag{3.3}$$

Clearly there are three possible rotational constants, denoted A, B and C. By convention, the constants are assumed to have the order $A \leq B \leq C$. As shown in the following subsections, these constants are used to classify the different types of possible molecular rotors.

Electric dipole-allowed rotational spectra generally require the molecule to possess a permanent dipole moment, usually denoted μ. This moment arises from the asymmetry of the charge distribution in the molecule. The intensity of the associated rotational transitions is proportional to $|\mu|^2$, so the magnitude of this dipole plays an important role in the observability of any transitions.

Pure rotational spectra obey the rather simple, general angular momentum selection rule $\Delta J = -1, (0), +1$, where $\Delta J = 0$ transitions can only occur non-trivially for systems with multiple energy levels for a given J. In practice, such transitions are important for asymmetric top molecules, see below.

The energy taken to excite a rotational level is relatively small meaning that transitions occur at long wavelengths. Many molecules show characteristic rotational spectra at radio frequencies while lighter molecules absorb and emit in the far infrared.

Rotational motions can be represented relatively easily: a basis of the $(2J + 1)$ rotational functions given by the Wigner rotation matrices are sufficient to characterise all levels for a given J. In practice parity can be used to symmetrise this basis into two sets of $(J + 1)$ and J functions. More information is given in specialist texts such as Bernath (2015).

In order to appreciate the rotational spectrum of different molecules it is necessary to first classify these species to different types of rotor. This is done on the basis of their moments of inertia, or, more usually, the rotational constants derived from them. This analysis shows that molecules can be classified into four classes of rotor. These are considered in turn below.

3.4.1 Linear Molecules

Linear molecules can only rotate about two axes and are therefore characterized by the rotational constants $A = 0$ and $B = C$. All diatomic molecules, including of course the abundant H_2 and CO molecules, are linear. A number of important polyatomic species such CO_2, HCN and HCCH are also linear.

The rotational energy levels of a rigid linear molecule are given by the formula

$$E_J = BJ(J + 1). \tag{3.4}$$

Using this formula with the $\Delta J = 1$ selection rule leads to the result that the rotational transitions are uniformly spaced in frequency/wavenumber/energy at intervals of $2B$. In practice molecules are non-rigid so this spacing actually decreases (slightly) with increasing J. However the uniform nature of the spectrum is easily recognised, see Fig. 3.5.

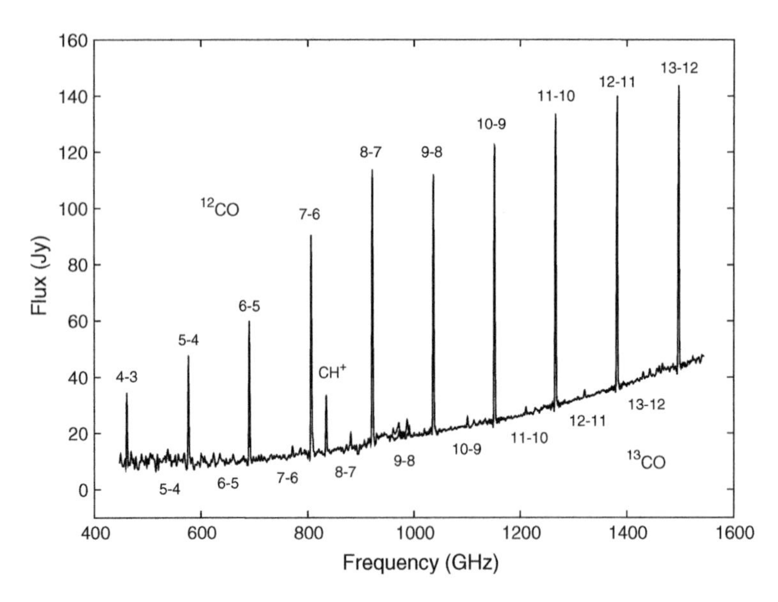

Fig. 3.5 Rotational spectrum of carbon monoxide (CO) recorded in emission from planetary nebula NGC 7027 using the using the SPIRE Fourier Transform Spectrometer on the Herschel Space Observatory. Note the even spacing between rotational transitions. (Adapted from R Wesson et al., *Astron. Astrophys.*, **518**, L144, 2010)

As already discussed, strong transitions require a dipole moment and for rotational transitions this is provided by the permanent dipole moment of the molecule. However, not all linear molecules have a permanent dipole. So-called homonuclear diatomic molecules such as H_2 and N_2 have zero dipole, as do symmetric linear polyatomic species such as CO_2 and HCCH. However, many important linear molecules do have a permanent dipole including all heteronuclear diatomic molecules such as CO and NaCl. Asymmetric linear polyatomic molecules such as HCN also have dipoles. As linear molecules concentrate all the intensity for transitions between neighbouring J levels into a single spectral line, their spectra tend to be strong. This is even true for the important CO molecule which actually possesses a rather small dipole moment.

The non-rigidity of the molecule leads to effects due to centrifugal distortion: the bond in the molecule stretches slightly as the molecule rotates. This leads to a small lowering in the rotational energy levels and a corresponding closing up of the transition frequencies. This effect can be represented as a power series expansion obtained using perturbation theory starting from the idea that rotations are rigid. Adding the leading correction to Eq. 3.4 gives:

$$E_J = BJ(J+1) - DJ^2(J+1)^2 + \cdots \tag{3.5}$$

which shows that centrifugal effects become much more pronounced at high J values.

Of course molecules also vibrate which means that rotational constants such as B and D also depend on the vibrational state of the molecule. This means that more generally expression (3.5) is written

$$E_J^v = B_v J(J+1) - D_v J^2(J+1)^2 + \cdots \tag{3.6}$$

where E_J^v represents the rotational energy of vibrational state v. The rotational constants B_v and D_v then also depend on the vibrational state with B_0 and D_0 characterizing the rotational energies of the ground (lowest) vibrational state. Rotational line positions both in the lab and astronomically can often be measured to high accuracy which involves extending the treatment given above. This is discussed in standard textbooks on molecular spectroscopy such as Bernath (2015).

Huber and Herzberg (1979) gave a comprehensive and very useful compilation of constants of diatomic molecules which is now available, unupdated, via the NIST chemistry web book (http://webbook.nist.gov/chemistry). This is a valuable, but dated resource.

3.4.2 Spherical Tops

Spherical top molecules are characterized by having all three rotational constant the same, $A = B = C$. This is the same property as any sphere and actually means that the orientation of the principle axes is arbitrary. The energy levels of spherical tops obey formulae similar to those for linear molecules discussed in Sect. 3.4.1 above. Astronomically important examples of spherical top molecules include methane (CH_4), silane (SiH_4) and buckminster fullerene (C_{60}).

The lack of any unique orientation for spherical top molecules means that they cannot possess a permanent dipole moment. This means that spherical tops should not undergo pure rotational transitions. In practice, as the molecule rotates it undergoes centrifugal distortion about the axis of rotation. This can induce a small, temporary dipole moment and leads to a weak rotational spectrum. Molecules containing hydrogen are easier to distort in this fashion and weak rotation spectra for methane are well-known, see Bray et al. (2017). However, they are likely to be too weak to be of great importance for exoplanetary spectroscopy.

3.4.3 Symmetric Tops

Symmetric top molecules have two moments of inertia which are identical and one which is different. They therefore occur in two distinct forms:

1. Prolate for which $A > B = C$. Prolate species can be thought of as egg-shaped.
2. Oblate for which $A = B > C$. Extreme oblate tops are flat like a disk. The planar symmetric molecule H_3^+ is an example of an oblate symmetric top.

The energy levels of a symmetric top are characterised by two quantum numbers, the usual J and K, which is the projection of J along the symmetry axis of the molecule. This extra quantum number lifts the $(2J + 1)$ degeneracy of the energy levels found for the rigid symmetric top. Since $|K| \leq J$, there are now $J + 1$ distinct levels. The level with $K = 0$ is singly degenerate and all other levels are two-fold degenerate.

The energy levels of a rigid prolate symmetric top can be written

$$E_J = BJ(J + 1) + (A - B)K^2, \tag{3.7}$$

while those for a rigid oblate symmetric top are

$$E_J = BJ(J + 1) + (C - B)K^2. \tag{3.8}$$

Transitions follow the general $\Delta J = 1$ selection with the strong propensity that K does not change, *i.e.* $\Delta K = 0$.

Most symmetric tops, including molecules like ammonia (NH_3) and phosphine (PH_3) have permanent dipole moments and hence rotational spectra. Figures 3.6 and 3.7 show room temperature rotational spectra of NH_3 and PH_3, respectively. The regular structure due to the $\Delta J = 1$ transition rule is clearly seen; the broadening of the features which can be seen at higher frequencies (high J) is due to the K substructure.

Planar symmetric-top molecules, such as H_3^+, do not possess a permanent dipole. However, H_3^+, like methane, is predicted to have a 'forbidden'rotational spectrum, see

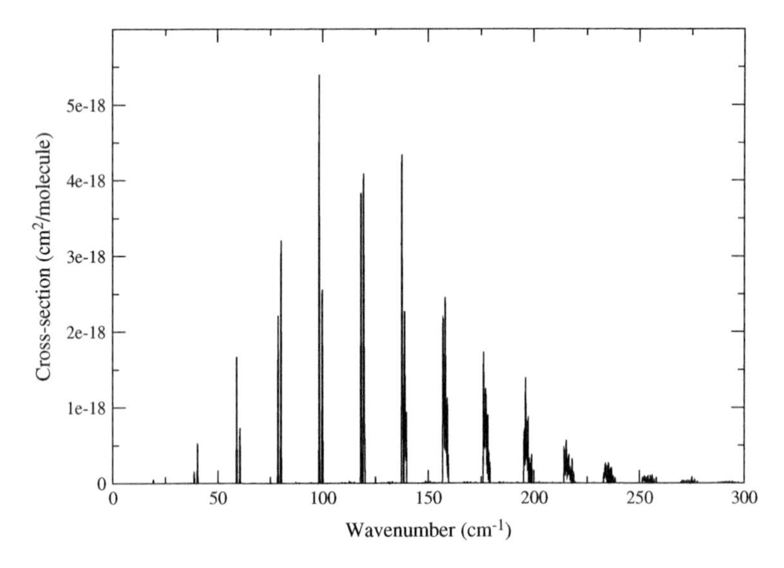

Fig. 3.6 Rotational spectrum of ammonia (NH_3) at 296 K generated using the BYTe line list of Yurchenko et al. (2011)

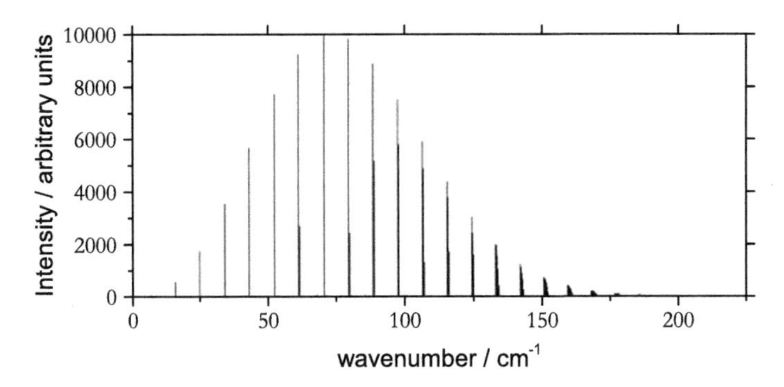

Fig. 3.7 Rotational spectrum of phosphine (PH_3) at 296 K generated using the line list of Sousa-Silva et al. (2013)

Miller and Tennyson (1988). This spectrum has yet to be observed in the laboratory or space but there is so much H_3^+ in environments such as Jupiter's aurora that it should be observable. However, the most prominent transitions lie in the far-infrared making them difficult to observe from Earth.

3.4.4 Asymmetric Tops

Asymmetric top molecules are characterized by having all moments of inertia different and in this case $A > B > C$. There are many asymmetric top species; common astronomical examples include water (H_2O) and formaldehyde (H_2CO). For both of these species it is possible to place the principle axes of rotation uniquely on symmetry grounds. For many molecules, such as ethanol (CH_3CH_2OH) this is not so and these axes plus their associated rotational constants have to be obtained by explicit diagonalisation of the moment of inertia tensor.

Asymmetric tops completely break the $(2J + 1)$-fold degeneracy of each J level found in symmetric tops and therefore require a new notation for the energy levels. In practice there are two notations. The more common one uses the projection of the rotational angular momentum, J, along the A and C axes giving approximate quantum numbers K_A and K_C, respectively. The states are generally denoted $J_{K_A K_C}$ e.g. 1_{10}. An alternative, more compact notation, uses τ which is defined as $\tau = K_A - K_C$. In this notation the $(2J + 1)$ levels run from $\tau = -J$ (lowest) to $\tau = +J$ (highest) and are denoted J_τ.

Even within the rigid rotor approximation the formula for the energy levels of an asymmetric top cannot in general be represented in a simple analytic form and the reader is referred to specialist books, such as Townes and Schawlow (2012), for more information.

The splitting of the energy levels results in asymmetric tops observing the more general selection rule that $\Delta J = 0, \pm 1$ meaning that transitions are allowed between different $K_A K_C$ within a given J. Although it is possible to find asymmetric tops without a permanent dipole, such as ethyne (CH_2CH_2) and ethane (CH_3CH_3), these are rare. Most asymmetric tops have a permanent dipole moment and complicated rotational spectra with little discernable pattern to them. A classic example is water whose spectrum spreads across large regions of the far infrared in a manner for which little or no systematic structure is readily apparent, see Fig. 3.8.

Rotational spectra of hot water can clearly be seen in a variety of locations. The spectrum of cold water is heavily obscured by water vapour in our own atmosphere and so cannot be observed from the ground. In contrast, however, much of the spectrum of hot water is shifted into the so-called water windows and so the spectra of hot water in astronomical sources can be observed from the ground. Figure 3.9 shows a small portion of a very detailed spectrum of a sunspot recorded from the Kitt Peak National Observatory.

3.4.5 Molecular Hydrogen

Molecular hydrogen, H_2, is a linear molecule with no permanent dipole moment. For rotational spectroscopy that should be the end of the matter but because the abundance of hydrogen in the Universe is so large it is necessary to consider all possibilities. Of course, gas giants are largely composed of hydrogen so its spectrum is very important for exoplanets.

Changing one H atom for a deuterium (D) creates an asymmetry. HD itself has a very small dipole moment of about 5×10^{-4} D, see Pachucki and Komasa (2008).

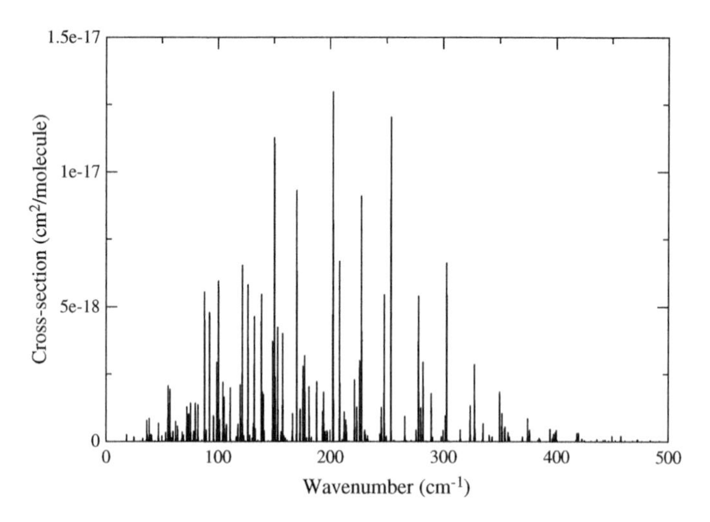

Fig. 3.8 Rotational spectrum of water at 296 K generated using the BT2 line list of Barber et al. (2006)

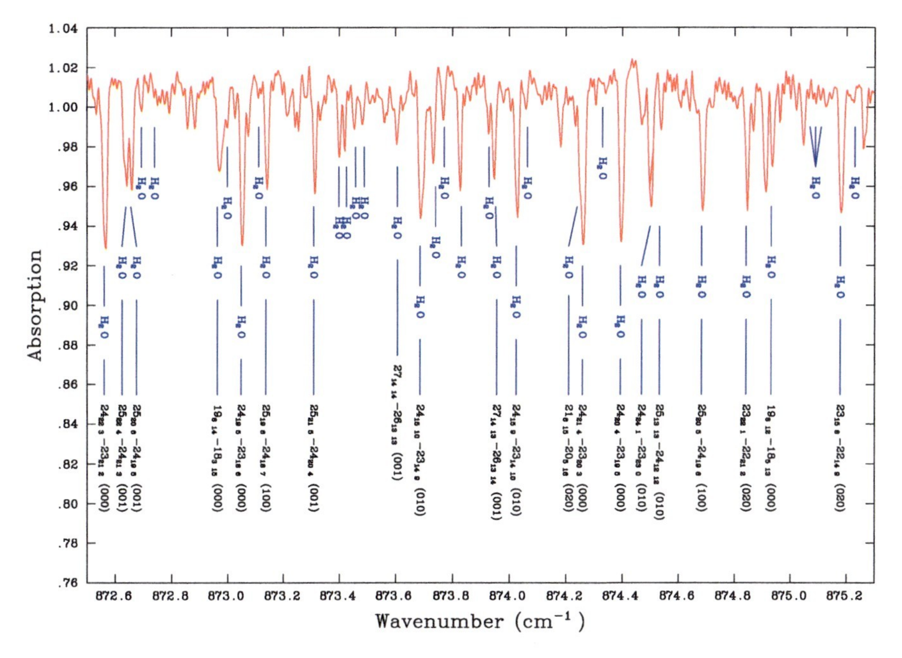

Fig. 3.9 Absorbtion spectrum of water recorded in sunspots by Wallace et al. (1996) with line assignments due to Polyansky et al. (1997)

The ionised form of HD, HD^+ has a much larger permanent dipole moment due to the separation between the centre-of-mass and centre-of-charge. A simple back-of-the-envelope calculation suggests that this should be about 0.85 D. As the transition intensity scales as the square of the dipole moment, HD^+ should be hugely easier to see than HD. However, so far the dipole-driven rotational spectrum of neither has been observed in space.

Turning to H_2 itself, this can be observed through very weak electric quadrupole transitions. Such transitions are about 10^8 times weaker than standard electric dipole transitions but clearly visible in space given the large columns of H_2. In fact it is much easier to see H_2 rotational transitions in space than in the lab!

Quadrupole transitions have the selection rule $\Delta J = 0, \pm 2$. Given that the rotational constant for H_2, B, equals 60.853 cm^{-1}, these transitions lie in the far infrared. In particular:

$J = 2 - 0$ lies at 28.22 μm;
$J = 3 - 1$ lies at 17.04 μm;
$J = 10 - 8$ lies at 5.05 μm.

These lines and many more have been observed in a variety of locations but not yet in exoplanets.

3.4.6 Isotopologues

Isotopologues are isotopically substituted molecules such as ^{13}CO for example. Rotational spectroscopy is the best way of detecting isotopologues and hence determining isotopic abundances. This is because there is a direct relationship between the rotational constant, B, and the mass of the atoms.

The rotational constant can be defined in terms of the moment of inertia, I, as:

$$B = \frac{\hbar^2}{2I} \tag{3.9}$$

The moment of inertia itself depends on the molecular (reduced) mass, μ, via the relationship

$$I = \mu R^2 \tag{3.10}$$

where R is the (effective) bondlength. For diatomic molecule such as CO the reduced mass is

$$\mu = \frac{M_C M_O}{M_C + M_O}. \tag{3.11}$$

This means changing from, for example, $^{12}C^{16}O$ to $^{13}C^{16}O$ gives a significant shift in line position:

$J = 1 - 0$: $\lambda(^{12}C^{16}O) = 2.60$ mm
$J = 1 - 0$: $\lambda(^{13}C^{16}O) = 2.72$ mm

Such shifts are straightforward to observe and, as a result, pure rotational spectra give the most sensitive method of studying astronomical isotope abundances.

3.4.7 Temperature Effects

Molecular spectra show strong temperature effects and can provide a useful thermometer in many astronomical environmemts. Except in the very coolest environments, molecules are usually found occupying a range of rotational energy levels. In thermodynamic equilibrium, the occupancy of a given level, P_i is determined by Boltzmann's distribution:

$$P_i = \sum_i \frac{g_i \exp\left(-\frac{E_i}{kT}\right)}{Q(T)} \tag{3.12}$$

where the sum runs over all levels, i, of energy E_i (with $E_0 = 0$) and degeneracy g_i. T is the temperature and k is Boltzmann's constant. $Q(T)$ is the partition function which is given by

$$Q(T) = \sum_i g_i \exp\left(-\frac{E_i}{kT}\right). \tag{3.13}$$

The partition function both keeps the probability of occupancy normalised and provides information on distribution of energy levels.

Partition functions are important for determining the behaviour of molecules as a function of temperature. At elevated temperatures one needs to be careful when computing them as the sum must run over many, may be all, levels in the system, see Vidler and Tennyson (2000). Gamache et al. (2017) give partition functions up to $T = 3000$ K for molecules occurring in the Earth's atmosphere. Barklem and Collet (2016) present partition functions for 291 diatomic molecules and atoms of astrophysical interest which updates but does not entirely replace earlier compilations by Irwin (1981) and Sauval and Tatum (1984). Finally the ExoMol database, see Tennyson et al. (2016c), contains partition functions that extend to high temperatures for all ExoMol molecules, see Table 3.6.

In the absence of a magnetic field, rotational states have degeneracy factor $g_J = (2J + 1)$ which means that the $J = 0$ state is not generally the most populated one. Instead the highest populated J state, $P_J^{(\max)}$, is given by the formula

$$P_J^{(\max)} = \frac{kT}{2B} - \frac{1}{2} \tag{3.14}$$

3.4.8 Data Sources

As mentioned above Huber and Herzberg (1979)[1] give data on rotational spectra of diatomic molecules. CDMS (The Cologne Database for Molecular Spectroscopy) (www.astro.uni-koeln.de/cdms) of Müller et al. (2005) and the JPL Molecular Spectroscopy database (spec.jpl.nasa.gov) due to Pickett et al. (1998) provide largely complementary information on spectra of a large range of astrophysically-important molecules at long wavelengths. As such the data are mainly concerned with the rotational spectra of these molecules. The virtual atomic and molecular data centre (VAMDC) provides a portal with access to a variety of data sources including CDMS and JPL, see Dubernet et al. (2016).

3.4.9 Other Angular Momentum

There are other sources of angular momentum in molecules. The full coupling can become complicated, see Brown and Carrington (2003). One of these sources of angular momentum is the spin of the nuclei. Because coupling between the rotational motion of the molecule and the nuclear spin is weak, the splitting of spectral lines due to it is small and is known as the hyperfine structure. The large splitting due to coupling to any electronic angular momentum is generally called fine structure.

[1] Available at http://webbook.nist.gov/chemistry.

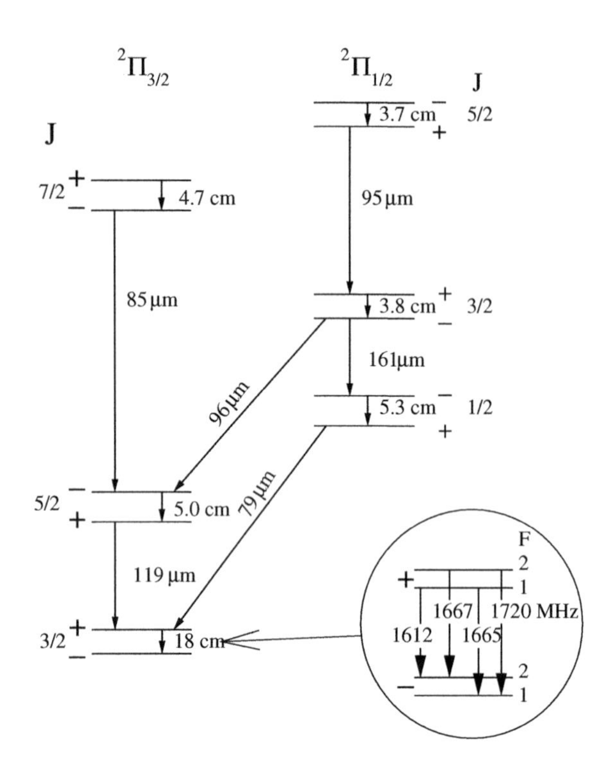

Fig. 3.10 Long-wavelength transitions between low-lying states in the OH molecule. The basic splitting of the rotational levels are due to spin-orbit coupling and represent fine structure. The expanded region shows the resolved hyperfine transitions within the lowest rotational transition

If the nuclear spin angular momentum is denoted I, the final angular momentum, F, can take values $F = |J - I|, |J - I + 1|, \ldots J + I - 1, J + I$. This hyperfine structure is quite often resolved in astronomical spectra; Lellouch et al. (2017) report such observation for HCN in the atmosphere of Pluto, for example. Remarkably, in some case it has been observed to be out of thermodynamic equilibrium.

Figure 3.10 shows the energy level splitting and long wavelength transition in the OH molecule illustrating both fine and hyperfine structures.

3.5 Vibrational Spectra

Vibrational spectra usually lie in the infrared and are important for the atmospheric properties of planets. Indeed, it is through the use of infrared vibration-rotation spectra (all vibrational spectra involve simultaneous changes in rotation) that nearly all detections of molecules in exoplanets have been made so far.

As discussed above, molecules vibrate with $3N - 6$ ($3N - 5$ for a linear molecule) degrees of freedom, which are generally termed vibrational modes. These vibrations represent generally small displacements from the equilibrium structure of the molecule.

The simplest assumption which works well in many cases is to assume that the vibrations are harmonic. The energy levels for the harmonic oscillator follow the rule:

$$E_v = \hbar\omega(v + \frac{1}{2}) \tag{3.15}$$

where v is the vibrational quantum number and ω is the angular frequency. In terms of molecular properties ω is given by

$$\omega = \left(\frac{k}{\mu}\right)^{\frac{1}{2}} \tag{3.16}$$

here k is the force constant as classically given by Hooke's law and μ is the reduced mass, which has already been discussed above for rotation.

Within the harmonic approximation, the vibrational modes are generally known as normal modes of vibration. Although changes of state by a single vibrational quantum usually gives the strongest transitions, other vibrational transitions are allowed and there is a standard terminology for describing these harmonics:

v_1: (1,0,0,...) – (0,0,0,...), the fundamental band;
nv_1: (n,0,0,...) – (0,0,0,...), overtone band, $n > 1$;
$nv_1 + mv_2$: (n,m,0,...) – (0,0,0,...), combination band;
$nv_1 - mv_1$: (n,0,0,...) – (m,0,0,...), hot band, $n > m > 0$;
$nv_1 - mv_2$: (n,0,0,...) – (0,m,0,...), difference bands.

These transitions are not subject to any rigorous selection rules on the change in vibrational quantum numbers. Generally changes with $\Delta v = 1$, such as in the fundamental, are much the strongest. But this is not always true: for example the important H_3^+ molecule has a very strong $\Delta v_2 = 2$ bending overtone which happens to lie in the K-band allowing it to be observed from ground-based telescopes; indeed the original spectroscopic detection of H_3^+ in Jupiter by Drossart et al. (1989) was based on a very strong K-band emission spectrum.

Vibrational transitions are usually accompanied by changes in the rotational quantum number. These changes mean that a vibrational transition gives a band with a characteristic structure rather than just a single line. For all bands, transitions which satisfy the selection rule $\Delta J = 1$ are allowed. $\Delta J = J' - J'' = -1$ transitions (where J' is the upper level and J'' the lower one) are called the P branch and often labelled P(J''). $\Delta J = +1$ transitions give the R branch with individual transitions labelled R(J''). In cases where $\Delta J = 0$ transitions are also allowed then the band also shows a Q branch which often manifests itself as a strong feature made up of many overlapping transitions near the band centre.

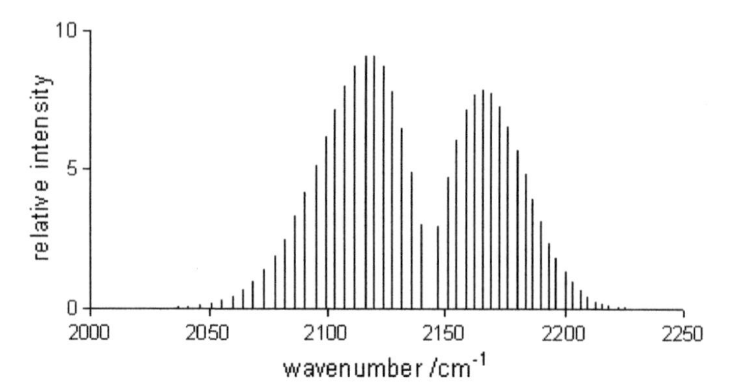

Fig. 3.11 The fundamental band of the CO molecule clearly showing the P-branch at frequencies below the band centre and R-branch at higher frequencies. The actual band centre is given by the missing line

3.5.1 Diatomic Molecules

Diatomic molecules only have one vibrational mode. Their spectra usually have a rather simple structure with a clearly visible P-branch at a lower frequency than the band centre and an R-branch at higher frequencies. Within the these two branches, transitions involving successive J levels are separated by approximately $2B$. In the absence of a Q-branch, the band centre is characterised by a gap of approximately $4B$ between the two branches. Figure 3.11 shows an example for the fundamental band of CO which clearly shows this structure.

The intensity of vibrational spectra relies on the change in the dipole moment upon vibrational excitation. Diatomics comprising two identical atoms, homonuclear diatomics, have zero dipole moment for internuclear separations so the dipole cannot change with vibrational change as it is always zero. These molecules do not have dipole allowed vibration-rotation spectra. All other diatomic molecules, heteronuclear diatomics, do have allowed spectra which, as discussed, generally lie in the mid-infrared although overtone bands can extend into the far-infrared.

3.5.2 Polyatomic Molecules

For molecules with more than two atoms it is necessary to specify the normal modes of vibration. Figure 3.12 illustrates this with the three normal modes for water.

Figure 3.12 also gives the vibrational frequencies, in the standard wavenumber units of cm^{-1}, for the fundamental modes of the water molecule. I note that $\omega_1 \approx \omega_3$ and $2\omega_2 \approx \omega_1 \approx \omega_3$. This observation has important consequences for the infrared spectrum of water as it means that bands associated with these motions overlap giving rise to what is know as polyad structures in the spectrum.

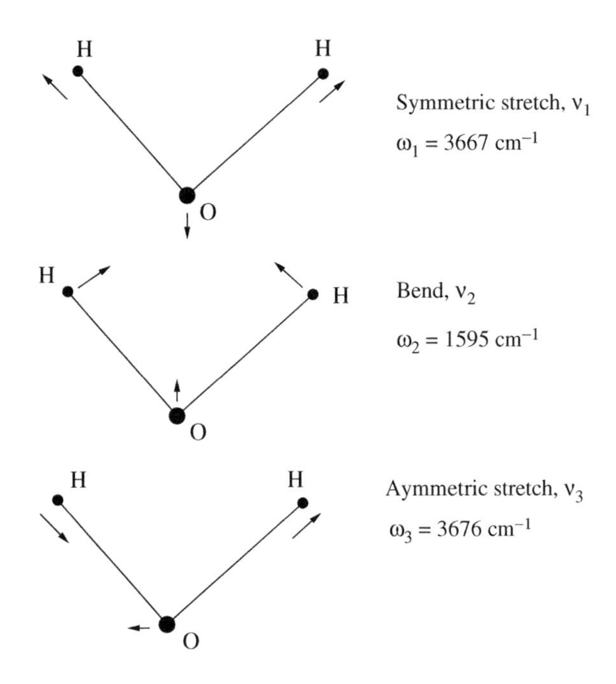

Fig. 3.12 Schematic representation of the water normal modes of vibration. Also given is the fundamental frequency of each mode

For water, the polyad number N of any vibrational state is given by $n_1 + \frac{1}{2}n_2 + n_3$ where n_1, n_2 and n_3 are the numbers of quanta of excitation of the respective vibrational modes. Polyad nomenclature varies from molecule to molecule; for water the polyad with $N = 1$ is referred to as either the first triad of 1δ and it contains three vibrational states, namely $(1, 0, 0)$, $(0, 2, 0)$ and $(0, 0, 1)$, where this notation is used to denote quanta of vibrational excitation, (n_1,n_2,n_3). The second polyad is like the first but has an extra quantum of excitation of the bending vibration, ν_2. This is known as the second triad or $1\delta + \nu$ and contains the three states $(1, 1, 0)$, $(0, 3, 0)$ and $(0, 1, 1)$. Water polyads are important not least because they are responsible for structure of absorbtion in the Earth's atmosphere with astronomical observing windows J, H, K, M. N and so forth lying in the gaps between these water bands.

Other molecules such as methane and ammonia also have a pronounced polyad structure. The structure of the methane polyads can clearly be seen in Fig. 3.16. This structure makes the rotation-vibration spectrum of methane appear fairly simple: this is not so. Each polyad is made up from a variety of different vibrational bands which, because they overlap, are hard or even impossible to disentangle. Understanding the observed spectrum of methane, particularly at near-infrared and red wavelengths, remains an active topic of research which is important for studies of exoplanets.

Species such H_2O, NH_3 and CH_4 all have vibrations involving hydrogen stretches (one or more H atoms vibrating against the central heavy atom). These stretches usually involve a significant change in the dipole moment and are therefore strong. They generally absorb or emit in the mid-infrared at wavenumbers 2000 to 3000

cm^{-1} (3–5 μm). Unfortunately, due to water absorbing very strongly in this region, these motions are hard to study from the ground.

All polyatomic molecules possess at least one "infrared active" vibrational mode. That is they all have at least one mode for which dipole allowed vibrational excitation can occur. This means that all polyatomic molecules can be detected by their infrared spectrum. For some molecules, such as CH_4, CO_2 and H_3^+ this probably represents the only practical way to observe the species astronomically. However, not all modes are infrared active. In the case of H_3^+ for example it is only the ν_2 bending mode which distorts the molecule to create an instantaneous dipole that has an allowed infrared spectra.

Rotation-vibration spectra of linear polyatomic molecules such as HCN or acetylene (HCCH) have a similar P-branch and R-brand structure to linear diatomics such as CO, see Fig. 3.11. For some bands there is also also a Q-branch corresponding to transitions where $J' = J''$; Q-branches manifest themselves by a strong peak near the band centre as all the Q-branch transitions lie at approximately the same wavenumber. Q-branch features can be used for detection of molecules at low resolution.

The rotation-vibration spectra of non-linear polyatomic molecules, particularly asymmetric tops, tend to have a much less pronounced structure. A classic example of this is the spectrum of water which is broad and unstructured; even transitions with $\Delta J = 0$ are spread out over a range of transition frequencies due to the substructure of the rotational levels.

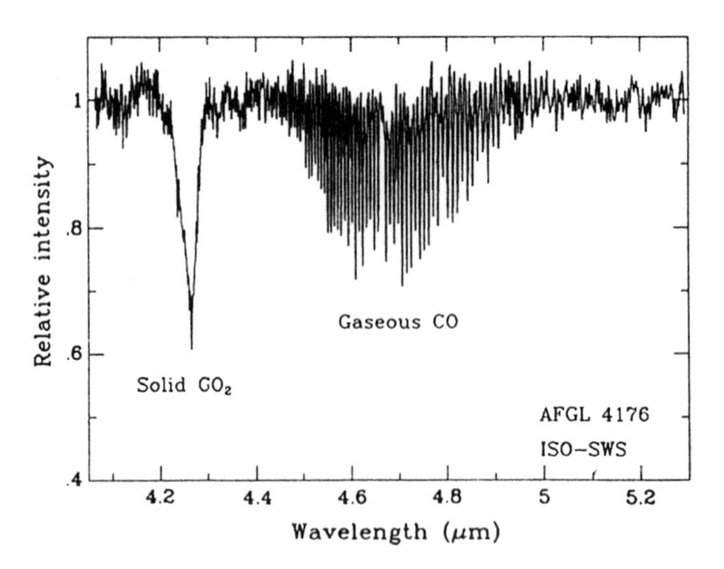

Fig. 3.13 Spectrum obtained with the Infrared Space Observatory toward the massive young stellar object AFGL 4176 in a dense molecular cloud. The strong, broad absorption at 4.27 μm is due to solid CO_2, whereas the structure at 4.4-4.9 μm indicates the presence of warm, gaseous CO along the line of sight. (Reproduced with permission from E.F. van Dishoeck, in *The Molecular Astrophysics of Stars and Galaxies*, eds. T.W. Hartquist and D.A. Williams (Clarendon Press, Oxford, 1998).)

Figure 3.13 shows an absorption spectrum of carbon monoxide recorded looking towards the massive young stellar object AFGL 4176, which is embedded in a dense molecular cloud. The shorter wavelength feature, which is due to solid carbon dioxide, shows how different these largely structureless condensed phase spectra are. The 4.7 μm region is heavily obscured by the Earth's atmosphere. This spectrum was recorded by a satellite, the Infrared Space Observatory (ISO).

3.5.3 *Isotopologues*

Isotopic substitution of molecules alters their rotation-vibration spectrum. As the vibrational frequency depends on the inverse square root of the molecular mass, see Eq. (3.16), these frequencies shift significantly upon isotopic substitution. These shifts can be resolved at reasonable resolution. Remember, however, that pure rotational spectra depend on the actual inverse of mass meaning that the shifts in the rotational transition frequencies are larger than those in rotation-vibration spectra.

There is one other important effect of isotopic substitution. Equation (3.15) shows that even in the $v = 0$ state there is a residual vibrational energy. This is known as the zero point energy and, like the transition frequencies, depends on the isotopic mass. This means isotopic substitution can change the energy of the system leading to an effect known as fractionation.

Consider the isotope exchange reaction:

$$D + H_2 \leftrightarrow H + HD. \tag{3.17}$$

The zero point energy of H_2, which is approximately 2200 cm^{-1}, exceeds that of HD, which is about 1905 cm^{-1}. This means that the reaction is exothermic by about 295 cm^{-1} or about 420 K. This means that at low temperatures the formation of HD is heavily favoured.

3.5.4 *Molecular Hydrogen*

H_2 has no permanent dipole moment so cannot undergo electric dipole-allowed vibration-rotation transitions.

However, there is a lot of hydrogen in the Universe. This means that very weak electric quadrupole transitions can occur instead. These are very weak: about a factor 10^{-8} weaker than dipole allowed transitions. The quadrupole transitions obey new selection rules: $\Delta J = +2$ which constitute the S-branch and $\Delta J = -2$ give the O-branch, as well as $\Delta J = 0$ transitions giving a Q-branch. H_2 transitions can be easily observed astronomically.

The infrared spectrum of Uranus is dominated by lines belonging to the fundamental ($v = 1 - 0$) rotation-vibration band of H_2. Figure 3.14 shows bright emission

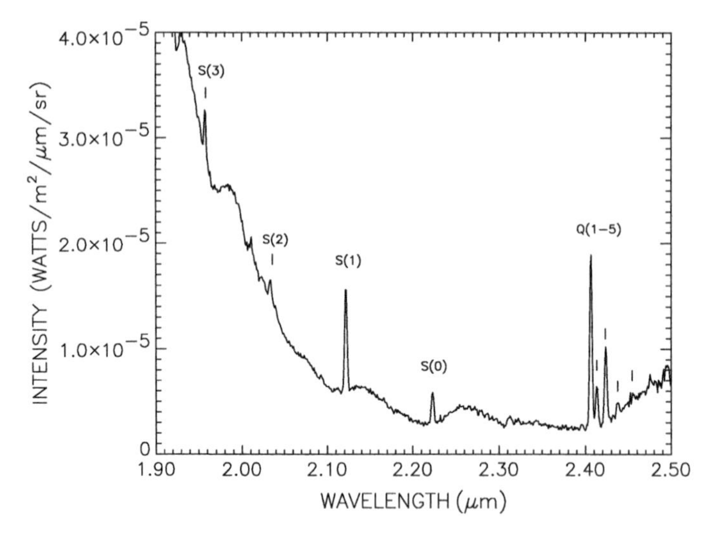

Fig. 3.14 K band infrared spectrum of Uranus showing H_2 quadrupole emission lines, adapted from Trafton et al. (1999)

lines, which are labeled, from both the S-branch ($\Delta J = +2$) and Q-branch ($\Delta J = 0$). The notation for assignments is S(J") and Q(J"), where J" is the rotational quantum number of the lower vibration state.

3.5.5 Temperature Effects

At room temperature most small molecules are predominantly in their vibrational ground state so spectra are dominated by transitions from this state. At higher temperatures, for example at temperatures found in the atmospheres of most exoplanets, higher vibrational states can become occupied. This leads to dramatic changes in the absorption spectrum.

Figure 3.15 compares the absorption spectrum of carbon monoxide at room temperature and at the temperature of a hot exoplanet.The figure shows the fundamental band about 2000 cm^{-1} plus the $v = 2 - 0$ and $v = 3 - 0$ overtone bands (note the standard notation is upper – lower). The cross section units are those adopted by the HITRAN data base of Gordon et al. (2017) and are macroscopic which means they require multiplication by Avogadro's number ($N_A = 6.022140857 \times 10^{23}$) to place them on a per molecule basis. The figure uses line-broadening parameters which results individual lines being blended into a single feature; however at 300 K the distinct P and R branches associated with each band can be seen. As the temperature is raised the peak of each band is reduced but the bands become significantly

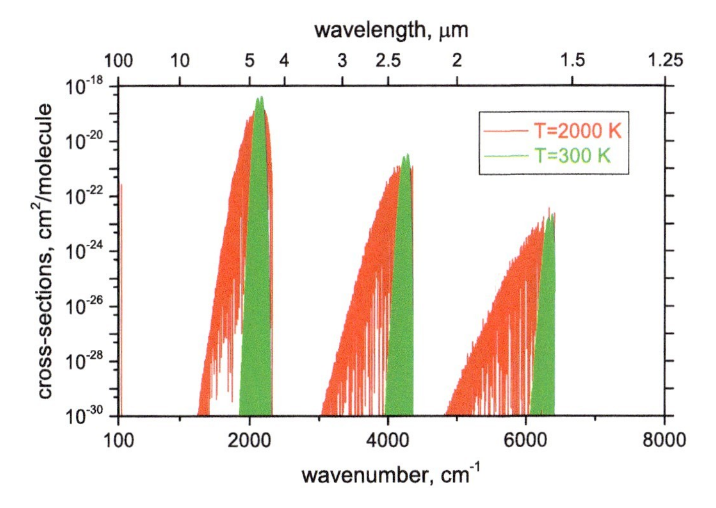

Fig. 3.15 Infrared absorption spectrum of carbon monoxide (CO) at 300 and 2000 K generated using the line list of Li et al. (2015)

broader. The integrated intensity of each band is roughly conserved as a function of temperature although it is difficult to judge this because of the log cross-section scale.

The broadening of the band with increasing temperature is caused by two effects. First more rotational levels are thermally occupied making the P and R branches broader. In practice the rotation-vibration spectrum of CO has a band head (see discussion of electronic spectra below) and this means that the broadening is largely to the red as temperature increases. At high resolution this particular feature of CO facilitates its use as thermometer in the atmospheres of cool stars and similar bodies, see Jones et al. (2005). Secondly at elevated temperatures excited vibrational states of CO become thermally occupied resulting in what are called hot band transitions of the type $v = 2 - 1$, $v = 3 - 1$ and $v = 4 - 1$. Anharmonic effects mean that the centres of these bands lie at slightly lower frequencies and again the band is largely red shifted. In summary as temperature is increased there is an increase in the number of observable transitions with the result that the band peaks are lowered and the band is broadened. In the case of CO the broadening shows a pronounced asymmetry with increased width towards the red.

Figure 3.16 gives a similar plot for methane. In this case there is a huge increase in the number of lines with temperature. Models of brown dwarfs demonstrated that it was necessary to include billions of spectral lines of methane alone to obtain reliable results, see Yurchenko et al. (2014). This is an important result which is of great significance for models of hot exoplanets.

Like that of CO, the effect of raising the temperature is to lower the peak absorption and to significantly broaden the bands. Unlike CO, at higher temperatures the windows between the bands disappear leading to the differences between the peaks

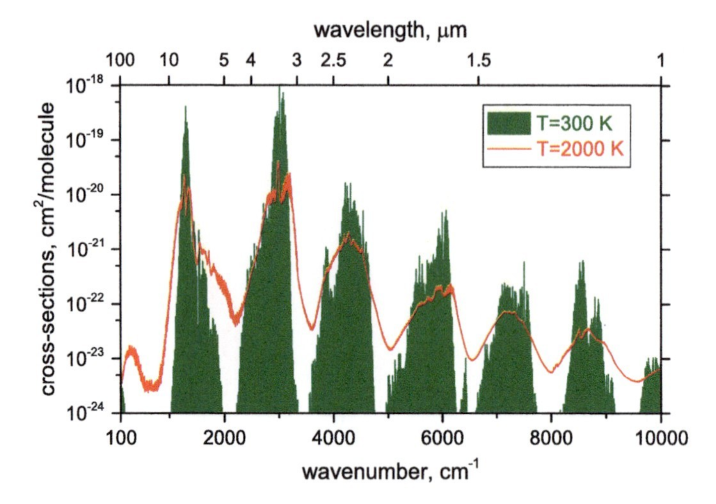

Fig. 3.16 Infrared absorption spectrum of methane (CH_4) at 300 and 2000 K generated using the 10 to 10 ExoMol line list of Yurchenko and Tennyson (2014)

and troughs being strongly flattened. This flattening is only achieved with very extensive line lists and is not given when, for example, a high temperature spectrum is modelled using a line list (such as those given by HITRAN) which are only complete for low temperatures. It was this realisation which led directly to the first successful detection of a molecule (water) in the atmosphere of a (hot Jupiter) exoplanet by Tinetti et al. (2007).

As for CO the lowering of the peaks and the flattening of the troughs in the hot methane spectrum is caused by a combination of high J rotational transitions and vibrational hot bands. In practice in polyatomic systems, starting from vibrationally excited states opens many new possibilities of transitions involving not only hot bands but also combination and difference bands. The net effect is to make both absorption and emission spectra of hot polyatomic molecules much flatter than the equivalent low temperature spectra. This has important consequences for both detection of these molecules, which is harder as the peaks are less pronounced, and radiative transport models as the opacity of hot molecules is widely distributed leading to blanket absorption. This can lead to trapping of radiation and significant increases in the size of atmospheres of bodies such as cool stars.

The large amount of data present in hot molecular line lists has led to treatments which lead to a more compact representation of the data. As the line list for a given molecule can be partitioned into strong and weak lines, the idea of representing the many, many weak lines using so-called super lines was suggested by Rey et al. (2017). An extensive discussion of this is given by Yurchenko et al. (2017).

3.6 Electronic Spectra

Electronic spectra involve a change of electronic state which corresponds to all spectra for atoms, which do not have the possibility of changes in vibrational or rotational motion. Important electronic spectra for standard closed-shell molecules such as all the molecules mentioned thus largely lie at ultra-violet wavelengths. However there is an important class of molecules which absorb at near-infrared or visible wavelengths. These are diatomic molecules containing a transition metal whose ground states are open shells. Open shells means that not all the electrons are paired and results in ground electronic states that are not ${}^1\Sigma^+$ states (see below for a discussion of this notation). Prominent examples of such species are TiO, VO, TiH and FeH. The spectrum of titanium monoxide (TiO) is particularly important as it is one of two major absorbers in the atmospheres of (cool) M-dwarf stars, the other important species is water (see Allard et al. 2000). These stars are the commonest ones in the local region of the Milky Way.

3.6.1 Electronc State Notation

For diatomic molecules the electronic state notation used for molecules has similarities to that used for atoms. The leading superscript is the electron spin degeneracy given as $2S+1$ where S is the electron spin, as for atoms. This is followed by a letter designating the projection of the orbital angular momentum along the molecular axis, Λ. Here the Greek letters $\Sigma, \Pi, \Delta, \Phi, \ldots$ representing $\Lambda = 0, 1, 2, 3, \ldots$, respectively, correspond to the atomic S, P, D, F, …notation for states with $L = 0, 1, 2, 3, \ldots$. This means that a ${}^1\Sigma^+$ state has no spin or orbital angular momentum as it corresponds to $S = \Lambda = 0$. This situation is referred to by chemists as a closed shell and most stable diatomics are closed shell species. Conversely a ${}^2\Pi$ state has both electron spin, $S = \frac{1}{2}$, and orbital angular momentum, $\Lambda = 1$.

As there are many states (actually an infinite number) of a given orbital-spin symmetry, each state is also given an additional, leading state label. X is used for ground electronic states e.g. X ${}^1\Sigma^+$. States with the same spin degeneracy as the ground state are labelled A, B, C, …from the lowest upwards, and those of different spin degeneracy are designated a, b, c, …. Unfortunately this notation is only applied in a rather haphazard fashion with a large number of, normally historical, anomalies.

Of course, besides an electronic state designation, each molecule state belongs to a particular vibrational and rotational state and so needs v and J quantum numbers to be fully specified.

Table 3.5 Selection rules for spectra of diatomic molecules undergoing allowed electric dipole transitions

Rotations	$\Delta J = \pm 1$	for $\Lambda = 0–0$,
	$\Delta J = 0, \pm 1$	not $J = 0–0$, for other $\Delta\Lambda$
Vibrations	Δv any (Franck-Condon approximation)	
Spin	$\Delta S = 0$	
Orbital	$\Delta\Lambda = 0, \pm 1.$	
Σ states	$\Sigma^+ \leftrightarrow \Sigma^+$,	
	$\Sigma^- \leftrightarrow \Sigma^-$	
Symmetry	$g \leftrightarrow u$	Homonuclear molecules only

3.6.2 *Selection Rules*

Selection rules for rotation-vibration-electronic transitions are given in Table 3.5. These transitions are often characterised by the portmanteau word rovibronic. Note that as transitions within an electronic state are explicitly allowed, the table also covers the cases of pure-rotational transitions and rotation-vibration transitions which are discussed above. The main rules for changes in electronic states for diatomic molecules are $\Delta S = 0$, $\Delta\Lambda = 0, \pm 1$. In addition, homonuclear diatomic molecules such as H_2 have an extra symmetry label according to whether the electronic state is symmetric (gerade or "g") or anti-symmetric (ungerade or "u"). This selection rule, which mirrors the Laporte rule for parity changes in atomic transitions, states that only g $\leftrightarrow$ u transitions are allowed.

There is no formal selection rule on vibrational transitions between different electronic states. Instead there are propensities which are controlled by the squared overlap of the vibrational wavefunctions in the upper and lower states. These terms, which are known as Franck-Condon factors, suggest that if two electronic states are approximately parallel then $\Delta v = 0$ transitions are strongly favoured. This is not the most common case and the reader is referred to Bernath (2015) for a more detailed discussion.

3.6.3 *Band Structure*

Rovibronic transitions like rotation-vibration ones obey the simple rotational selection that $\Delta J = 0, \pm 1$. However even for diatomic molecules, this may not lead to simple structures in the spectrum. This because the bondlength in the two electronic states involved in the transitions usually differ, resulting in very different rotational constants, see Eq. (3.10). This means that instead of the rotational structure producing a clear regular structure, the band tends to close-up in either the P or R branch resulting in the formation of a sharp feature known as a band head. Figure 3.17 shows

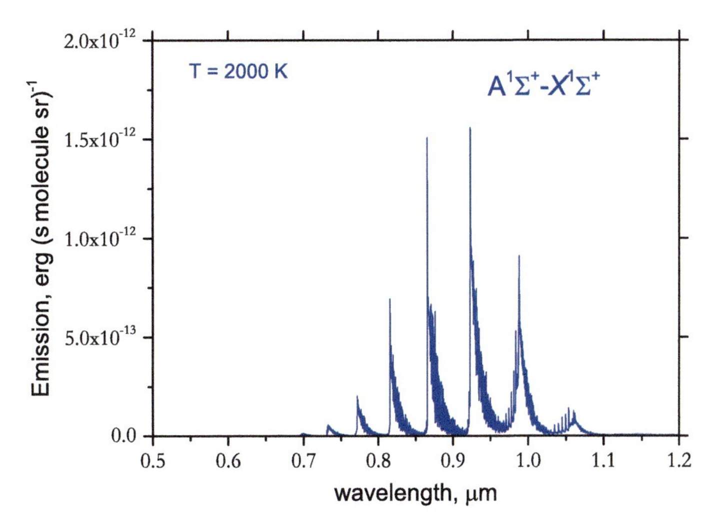

Fig. 3.17 Emission spectrum from the CaO A $^1\Sigma^+$ – X $^1\Sigma^+$ band of CaO at a temperature of 2000 K generated using the VBATHY ExoMol line list of Yurchenko et al. (2016a)

an emission spectrum for CaO; the spectrum shows clear band heads. These band heads can provide molecular markers even at relatively low resolution.

3.6.4 Duo

For exoplanet studies it has so far been found that electronic spectra of importance belong to diatomic species. Computing the spectra of diatomics from given potential energy curves is in principle straightforward. For closed shell species the program Level due to Le Roy (2017) has long provided an excellent means of computing spectra.

However, the spectra of open shell molecules are significantly more complex due to need to consider the various angular momentum couplings that arise between the rotational, spin and electronic angular momentum, see Tennyson et al. (2016b) for a comprehensive discussion of these. Yurchenko et al. (2016b) developed a new general diatomic code called Duo for computing rovibronic spectra of diatomic molecules and based on a rigorous treatment of these couplings in diatomic systems. Duo is freely available from the CCPforge program depository (ccpforge.cse.rl.ac.uk).

3.6.5 Molecular Hydrogen

H_2 has a number of strong electronic bands. The most prominent are the Werner (C $^1\Pi_u - X\,^1\Sigma_g^+$) and the Lyman (B $^1\Sigma_u^+ - X\,^1\Sigma_g^+$) bands. Note that the H_2 Lyman band is distinct from H atom Lyman lines. These bands are strong but lie in ultraviolet regions which means that they can only be observed from space.

3.7 Line Profiles

Spectral lines do not occur as a completely sharp feature at a unique frequency, instead each line has a profile. This profile is made up of a number of contributions:

1. Natural broadening: because each excited state has a finite lifetime, the uncertainty principle dictates that it must also have a finite spread (uncertainty) in its energy. This leads to a Lorentzian line profile. However, natural broadening is generally very small and is therefore only significant for the most precise laboratory measurements, Natural broadening is normally ignored for astronomical observations.
2. Doppler broadening: due to thermal motion of the molecules. This is distinct from Doppler shifts such as the red-shift which are due to overall motion of the body of gas being observed. Doppler broadening gives a Gaussian line profile which depends only on the (translational) temperature of the species concerned and its mass.
3. Pressure broadening: molecules in an atmosphere experience constant collisions due to the bombardment by other species present in the atmosphere. These collisions leads to small changes to the environment in which a transition occurs and hence to both pressure broadening of the transition and small shifts in the line position. The effect on the line shape can approximately be represented by a Lorentzian profile. The width of this profile of course depends on the pressure but also nature of the colliding species and the quantum numbers of the states involved in the transition. In practice, it is found for molecules that pressure-broadening depends more strongly on rotational state than on the precise vibrational states involved. In particular pressure effects decrease with rotational excitation. Pressure can also lead to mixing between lines. Line-mixing effects are particularly important when transitions lie close together, such as in Q-branches. However, this process can probably be ignored in most astronomical applications.
4. Other effects: lines are also broadened and shifted in the presence of electron collisions, this is particularly true for ions for which the interactions are long-range. As discussed above, magnetic fields can lead to a splitting in spectral lines: if the field is weak or the splitting is not resolved, this can also appear as a line broadening effect.

The most important effects in planetary atmospheres are due to Doppler and pressure broadening. As these effects depend respectively on Gaussian and Lorentzian profiles, the two effects need to be convolved. The standard convolution is the so-called Voigt profile and this is the one routinely used in models of planetary atmospheres. The Voigt profile is not analytic but there are a number of procedures available for evaluating it.

The pressure part of the Voigt profile also requires data for each transition and broadening. There is a severe lack of reliable information on broadening by key species such as H_2 and He, particularly at the elevated temperatures found, for example, in hot Jupiter exoplanets.

The effect of using a Voigt profile is to re-distribute the net absorption but, if the line is optically thin, the overall flux absorbed does not change. This is not true for optically thick lines for which the degree of broadening is crucial in determining the net absorption. Given that observations of exoplanets in primary transit give rise to long atmospheric pathlengths across the limb of the planet, it is inevitable that at least some of the lines will become optically thick. The importance of broadening under these circumstances has been demonstrated by Tinetti et al. (2012), who showed that pressure effects were particularly strong for long wavelength transitions.

In practice the Voigt profile is only an approximation to the true line profile as there are a number of subtle collisional effects which need to be considered to fully reproduce observed line-shapes, see Tennyson et al. (2014). Both Hartmann et al. (2008) and Buldyreva et al. (2010) provide comprehensive textbooks on line broadening to which the reader is referred for more information.

3.8 Spectroscopic Database

Access to data is an important issue for astronomical models. This section considers some useful sources of data. In principle the virtual atomic and molecular data centre (VAMDC) portal of Dubernet et al. (2016) provides access to the key spectroscopic data (and useful collision data) for astrophysical problems. However many of the datasets needed for molecular spectroscopy are too large to be handled by VAMDC; in this case it is necessary to go to the actual database.

3.8.1 Ground Rules

Before considering data sources in detail, it is worth considering some basic rules of what data to use and how to use them.

The HITRAN database, discussed further below, has been honed for many years to meet the needs of the large scientific community interested in studies of the Earth's atmosphere. It is an excellent source of data but explicitly designed to work at Earth temperatures. Indeed the data representation is all based on a reference temperature of 296 K. The nature of spectroscopic data is such that HITRAN data are probably suitable for modelling Earth-like objects at lower temperatures. However, at higher temperatures molecular spectra involve many more lines than are in HITRAN and use of the line lists provided by HITRAN will lead to spectroscopic features which are both systematically missing flux and, perhaps even more importantly, have the wrong band profiles.

Modellers should follow the following simple rules:

1. When modelling an Earth-like planet at temperatures below about 400 K HITRAN provides and excellent source of data.
2. When modelling temperatures above 400 K do not use HITRAN data.
3. When modelling data for hydrogen-rich systems such as gas giants data have to be taken from sources other than HITRAN, which does not aim to complete even for cool systems which are not oxygen-rich, such as Titan.
4. Do not claim that data come from HITRAN when they do not! This point may seem obvious but unfortunately there are frequent occurrence in the literature of this happening.
5. Equally obvious, but also ignored: cite the data sources used.

3.8.2 Atomic Data Sources

The NIST Atomic Spectra Database (https://www.nist.gov/pml/atomic-spectra-database) provides an excellent source of atomic spectroscopic data. It contains detailed listings of atomic energy levels (term values) and Einstein A coefficients (transition probabilities). In addition the Vienna atomic line database (VALD), see Ryabchikova et al. (2015), TOPbase (Cunto et al. 1993) and the extensive compilations due to Kurucz (2011) provide comprehensive data on ions important for stellar models.

3.8.3 The ExoMol Project

The ExoMol project was started by Tennyson and Yurchenko (2012) with the explicit aim of providing comprehensive molecular line lists for exoplanet and other atmospheres, Table 3.6 lists the currently available ExoMol molecular line lists. Table 3.7 list other line lists also available from the ExoMol website (www.exomol.com).

The ExoMol database, as described by Tennyson et al. (2016c), contains a number of features useful for models and other activities:

1. The database has a fully-configured applications program interface (API) which allows data, which can be dynamically updated, to be read directly into an application program.
2. Cross sections as function of temperature are presented for zero pressure and in many cases for H_2/He atmospheres.
3. Partition functions valid up to high temperatures are presented.
4. Pressure broadening parameters for H_2 and He as a function of temperature and rotational quantum are presented using the so-called ExoMol diet of Barton et al. (2017).
5. Tables of k-coefficients are presented for key species.

Table 3.6 Datasets created by the ExoMol project and included in the ExoMol database

Paper	Molecule	N_{iso}	T_{max}	N_{elec}	N_{lines}	DSName	Reference
I	BeH	1	2000	1	16 400	Yadin	Yadin et al. (2012)
I	MgH	3	2000	1	10 354	Yadin	Yadin et al. (2012)
I	CaH	1	2000	1	15 278	Yadin	Yadin et al. (2012)
II	SiO	5	9000	1	254 675	EJBT	Barton et al. (2013)
III	HCN/HNC	2[a]	4000	1	399 000 000	Harris	Barber et al. (2014)
IV	CH_4	1	1500	1	9 819 605 160	YT10to10	Yurchenko and Tennyson (2014)
V	NaCl	2	3000	1	702 271	Barton	Barton et al. (2014)
V	KCl	4	3000	1	1 326 765	Barton	Barton et al. (2014)
VI	PN	2	5000	1	142 512	YYLT	Yorke et al. (2014)
VI	PH_3	1	1500	1	16 803 703 395	SAlTY	Sousa-Silva et al. (2015)
VIII	H_2CO	1	1500	1	10 000 000 000	AYTY	Al-Refaie et al. (2015)
IX	AlO	4	8000	3	4 945 580	ATP	Patrascu et al. (2015)
X	NaH	2	7000	2	79 898	Rivlin	Rivlin et al. (2015)
XI	HNO_3	1	500	1	6 722 136 109	AIJS	Pavlyuchko et al. (2015)
XII	CS	8	3000	1	548 312	JnK	Paulose et al. (2015)
XIII	CaO	1	5000	5	21 279 299	VBATHY	Yurchenko et al. (2016a)
XIV	SO_2	1	2000	1	1 300 000 000	ExoAmes	Underwood et al. (2016b)
XV	H_2O_2	1	1250	1	20 000 000 000	APTY	Al-Refaie et al. (2016)
XIV	H_2S	1	2000	1	115 530 3730	AYTY	Azzam et al. (2016)
XV	SO_3	1	800	1	21 000 000 000	UYT2	Underwood et al. (2016a)
XVI	VO	1	2000	13	277 131 624	VOMYT	McKemmish et al. (2016)
XIX	$H_2{}^{17,18}O$	4[b]	3000	1	1 500 000 000	HotWat78	Polyansky et al. (2017)
XX	H_3^+	2[c]	3000	1	11 500 000 000	MiZATeP	Mizus et al. (2017)
XXI	NO	6	5000	2	2 281 042	NOName	Wong et al. (2017)
XXII	SiH_4	1	1200	1	62 690 449 078	OY2T	Owens et al. (2017a)
XXIII	PO	1	3000	1	1 500 000 000	POPS	Prajapat et al. (2017)
XXIII	PS	1	3000	3	1 500 000 000	POPS	Prajapat et al. (2017)
XXIV	SiH	4	5000	3	1 724 841	SiGHTLY	Yurchenko et al. (2018a)

N_{iso} Number of isotopologues considered;
T_{max} Maximum temperature for which the line list is complete;
N_{elec} Number of electronic states considered;
N_{lines} Number of lines: value is for the main isotope.
DSName Data set name for that line list.
[a] A line list for $H^{13}CN/HN^{13}C$ due to Harris et al. (2008) is also available.
[b] Line lists for $H_2{}^{16}O$ (BT2) due to Barber et al. (2006) $HD^{16}O$ (VTT) due to Voronin et al. (2010) are also available.
[c] A line list for H_2D^+ due to Sochi and Tennyson (2010) is also available

6. State-resolved lifetimes are presented for each species, see Tennyson et al. (2016a). These lifetimes can be used to obtain critical densities amongst other things.
7. Temperature-dependent cooling functions are presented.

Table 3.7 Other molecular line lists which can be obtained from the ExoMol website

Molecule	N_{iso}	T_{max}	N_{elec}	N_{lines}	DSName	Reference	Methodology
H_2O	2[a]	3000	1	505 806 202	BT2	Barber et al. (2006)	ExoMol
NH_3	2[b]	1500	1	1 138 323 351	BYTe	Yurchenko et al. (2011)	ExoMol
HeH^+	4	10000	1	1 431	Engel	Engel et al. (2005)	Ab initio
HD^+	1	12000	1	10 119	CLT	Coppola et al. (2011)	Ab initio
LiH	1	12000	1	18 982	CLT	Coppola et al. (2011)	Ab initio
LiH^+	1	12000	1	332	CLT	Coppola et al. (2011)	Ab initio
ScH	1	5000	6	1 152 827	LYT	Lodi et al. (2015)	Ab initio
MgH	1		3	30 896	13GhShBe	GharibNezhad et al. (2013)	Empirical
CaH	1		2	6000	11LiHaRa	Li et al. (2012)	Empirical
NH	1		1	10 414	14BrBeWe	Brooke et al. (2014a)	Empirical
CH	2		4	54 086	14MaPlVa	Masseron et al. (2014)	Empirical
CO	9	9000	1	752 976	15LiGoRo	Li et al. (2015)	Empirical
OH	1	6000	1	45 000	16BrBeWe	Brooke et al. (2016)	Empirical
CN	1		1	195 120	14BrRaWe	Brooke et al. (2014b)	Empirical
CP	1		1	28 735	14RaBrWe	Ram et al. (2014)	Empirical
HCl	1		1	2588	11LiGoBe	Li et al. (2011)	Empirical
CrH	1		2	13 824	02BuRaBe	Burrows et al. (2002)	Empirical
FeH	1		2	93 040	10WeReSe	Wende et al. (2010)	Empirical
TiH	1		3	181 080	05BuDuBa	Burrows et al. (2005)	Empirical

N_{iso} Number of isotopologues considered;
T_{max} Maximum temperature for which the line list is complete;
N_{elec} Number of electronic states considered;
N_{lines} Number of lines: value is for the main isotope.
[a]The VTT line list for HDO due to Voronin et al. (2010) is also available.
[b]There is a room temperature $^{15}NH_3$ line list due to Yurchenko (2015)

8. Landé g-factors are given for open shell diatomic molecules, see Semenov et al. (2017),
9. Finally, transition dipoles are stored; these are being used for studies of molecular control/orientation effects, aee Owens et al. (2017b). In future this information could also be useful for studies of polarisation effects.

In addition the program ExoCross of Yurchenko et al. (2018b) can be used to process line lists and to convert between ExoMol and HITRAN formats. Data can also converted to the Phoenix format of Jack et al. (2009), which is widely used for brown dwarf and other models.

3.8.4 *Other Data Sources*

As discussed above the HITRAN database contains comprehensive sets of spectroscopic parameters. The 2016 HITRAN release of Gordon et al. (2017) contains line-by-line data for 49 molecules of importance to models of the Earth's atmosphere plus a very extensive set of cross sections for larger molecules. In particular these cross sections have subsumed the PNNL library of Sharpe et al. (2004). HITRAN data are designed for use at terrestrial temperatures; the HITEMP database of Rothman et al. (2010) is designed to extend HITRAN to higher temperatures. However, HITEMP only contains data for five molecules: water, CO_2, CO, OH and NO. Newer and improved line lists are available for these species making HITEMP essentially redundant; release of a new edition of HITEMP is overdue.

The GEISA database of Jacquinet-Husson et al. (2016) contains data on 52 species. GEISA is very similar in design and content to HITRAN although it does contain some line data on carbon-containing molecules not included in HITRAN.

The compilations by Kurucz (2011) contains some data on diatomic molecules largely based on use of the diatomic constants given by Huber and Herzberg (1979). These data are not particularly accurate but are useful when no other data are available. Similarly the recent release of VALD (Ryabchikova et al. 2015) contains line list for a few diatomics as generated by Plez.

Other data sources includes the TheoReTS database of Rey et al. (2016) which is particularly useful for hot hydrocarbon species. The group at NASA Ames have generated extensive datasets for hot molecules found in the atmospheres of Venus and Mars, including recent line lists for all major isotopologues of CO_2 (Huang et al. 2017). Bernath and co-workers have assembled extensive empirical line lists for a variety of molecules, most of these are available via the ExoMol website, see Table 3.7.

3.9 Conclusion

To understand the spectroscopic signatures of atoms and molecules in the atmospheres of planets and elsewhere requires a wealth of spectroscopic data. This chapter gives a brief introduction to the physics involved in the various spectroscopic processes encountered in atoms and molecules. Of necessity many important issues are dealt with only briefly or not at all. However, I have tried to point to both sources of further information on the topic and useful sources of spectroscopic data for studies of exoplanets. I would like to end with one more comment. Any model is only as good as the data input into it. The use of good laboratory data is therefore crucial. Equally important when presenting results is to provide the source of these data. Not to do so is both to make any results presented unreproducible and to cast doubt on the validity of any results because of the unknown provenance of the data employed. In short: please cite your data sources.

Acknowledgements I thank the members of the ExoMol team for their contribution to the work reported here; in particular I would like to thank Sergey Yurchenko both for his contribution to the project and his help with preparing some of the figures shown here. I also thank Phillip Coles and Clara Sousa-Silva for supplying figures, and Tony Lynas-Gray for many helpful comments on the original manuscript.

References

Al-Refaie, A.F., Yurchenko, S.N., Yachmenev, A., Tennyson, J.: MNRAS **448**, 1704 (2015)
Al-Refaie, A.F., Polyansky, O.L., Ovsyannikov, R.I., Tennyson, J., Yurchenko, S.N.: MNRAS **461**, 1012 (2016)
Allard, F., Hauschildt, P.H., Schwenke, D.: ApJ **540**, 1005 (2000)
Azzam, A.A.A., Yurchenko, S.N., Tennyson, J., Naumenko, O.V.: MNRAS **460**, 4063 (2016)
Barber, R.J., Tennyson, J., Harris, G.J., Tolchenov, R.N.: MNRAS **368**, 1087 (2006)
Barber, R.J., Strange, J.K., Hill, C., Polyansky, O.L., Mellau, G.C., Yurchenko, S.N., Tennyson, J.: MNRAS **437**, 1828 (2014)
Barklem, P.S., Collet, R.: A&A **588**, A96 (2016)
Barton, E.J., Yurchenko, S.N., Tennyson, J.: MNRAS **434**, 1469 (2013)
Barton, E.J., Chiu, C., Golpayegani, S., Yurchenko, S.N., Tennyson, J., Frohman, D.J., Bernath, P.F.: MNRAS **442**, 1821 (2014)
Barton, E.J., Hill, C., Czurylo, M., Li, H.Y., Hyslop, A., Yurchenko, S.N., Tennyson, J.: J. Quant. Spectrosc. Radiat. Transf. **203**, 490 (2017)
Berdyugina, S.V., Solanki, S.K.: A&A **365**, 701 (2002)
Berdyugina, S.V., Braun, P.A., Fluri, D.M., Solanki, S.K.: A&A **444**, 947 (2005)
Bernath, P.F.: Spectra of Atoms and Molecules, 3rd edn. Oxford University Press (2015)
Bray, C., Cuisset, A., Hindle, F., Mouret, G., Bocquet, R., Boudon, V.: J. Quant. Spectrosc. Radiat. Transf. **203**, 349 (2017)
Brooke, J.S.A., Bernath, P.F., Western, C.M., van Hemert, M.C., Groenenboom, G.C.: J. Chem. Phys. **141**, 054310 (2014a)
Brooke, J.S.A., Ram, R.S., Western, C.M., Li, G., Schwenke, D.W., Bernath, P.F.: ApJS **210**, 23 (2014b)
Brooke, J.S.A., Bernath, P.F., Western, C.M., Sneden, C., Afsar, M., Li, G., Gordon, I.E.: J. Quant. Spectrosc. Radiat. Transf. **138**, 142 (2016)
Brown, J.M., Carrington A.: Rotational Spectroscopy of Diatomic Molecules. Cambridge University Press (2003)
Buldyreva, J., Lavrentieva, N., Starikov, V.: Collisional Line Broadening and Shifting of Atmospheric Gases: A Practical Guide for Line Shape Modelling by Current Semi-classical Approaches. Imperial College Press, London (2010)
Burrows, A., Ram, R.S., Bernath, P., Sharp, C.M., Milsom, J.A.: ApJ **577**, 986 (2002)
Burrows, A., Dulick, M., Bauschlicher, C.W., Bernath, P.F., Ram, R.S., Sharp, C.M., Milsom, J.A.: ApJ **624**, 988 (2005)
Cauley, P.W., Redfield, S., Jensen, A.G.: ApJ **153**, 81 (2017)
Coppola, C.M., Lodi, L., Tennyson, J.: MNRAS **415**, 487 (2011)
Cunto, W., Mendoza, C., Ochsenbein, E., Zeippen, C.J.: A&A **275**, L5 (1993)
Dello Russo, N., Bonev, B.P., DiSanti, M.A., Gibb, E.L., Mumma, M.J., Magee-Sauer, K., Barber, R.J., Tennyson, J.: ApJ **621**, 537 (2005)
Drossart, P., Maillard, J.P., Caldwell, J., et al.: Nature **340**, 539 (1989)
Dubernet, M.L., Antony, B.K., Ba, Y.A., et al.: J. Phys. B: At. Mol. Opt. Phys. **49**, 074003 (2016)
Engel, E.A., Doss, N., Harris, G.J., Tennyson, J.: MNRAS **357**, 471 (2005)
Gamache, R.R., Roller, C., Lopes, E., et al.: J. Quant. Spectrosc. Radiat. Transf. **203**, 70 (2017)
GharibNezhad, E., Shayesteh, A., Bernath, P.F.: MNRAS **432**, 2043 (2013)
Gordon, I.E., Rothman, L.S., Babikov, Y., et al.: J. Quant. Spectrosc. Radiat. Transf. **203**, 3 (2017)
Harris, G.J., Larner, F.C., Tennyson, J., Kaminsky, B.M., Pavlenko, Y.V., Jones, H.R.A.: MNRAS **390**, 143 (2008)
Hartmann, J.M., Boulet, C., Robert, D.: Collisional effects on molecular spectra. Laboratory experiments and models, consequences for applications. Elsevier, Amsterdam (2008)
Huang, X., Schwenke, D.W., Freedman, R.S., Lee, T.J.: J. Quant. Spectrosc. Radiat. Transf. **203**, 224 (2017)

Huber, K.P., Herzberg, G.: Molecular Spectra and Molecular Structure IV. Constants of Diatomic Molecules. Van Nostrand Reinhold Company, New York (1979)
Irwin, A.W.: ApJS **45**, 633 (1981)
Jack, D., Hauschildt, P.H., Baron, E.: A&A **502**, 1043 (2009)
Jacquinet-Husson, N., Armante, R., Scott, N.A., et al.: J. Mol. Spectrosc. **327**, 31 (2016)
Jones, H.R.A., Pavlenko, Y., Viti, S., Barber, R.J., Yakovina, L., Pinfold, D., Tennyson, J.: MNRAS **358**, 105 (2005)
Kurucz, R.L.: Can. J. Phys. **89**, 417 (2011)
Le Roy, R.J.: J. Quant. Spectrosc. Radiat. Transf. **186**, 167 (2017)
Lellouch, E., Gurwell, M., Butler, B., et al.: Icarus **286**, 289 (2017)
Li, G., Gordon, I.E., Bernath, P.F., Rothman, L.S.: J. Quant. Spectrosc. Radiat. Transf. **112**, 1543 (2011)
Li, G., Harrison, J.J., Ram, R.S., Western, C.M., Bernath, P.F.: J. Quant. Spectrosc. Radiat. Transf. **113**, 67 (2012)
Li, G., Gordon, I.E., Rothman, L.S., Tan, Y., Hu, S.M., Kassi, S., Campargue, A., Medvedev, E.S.: ApJS **216**, 15 (2015)
Lodi, L., Yurchenko, S.N., Tennyson, J.: Mol. Phys. **113**, 1559 (2015)
Masseron, T., Plez, B., Van Eck, S.: A&A **571**, A47 (2014)
McKemmish, L.K., Yurchenko, S.N., Tennyson, J.: MNRAS **463**, 771 (2016)
Miller, S., Tennyson, J.: ApJ **335**, 486 (1988)
Mizus, I.I., Alijah, A., Zobov, N.F., Kyuberis, A.A., Yurchenko, S.N., Tennyson, J., Polyansky, O.L.: MNRAS **468**, 1717 (2017)
Müller, H.S.P., Schlöder, F., Stutzki, J., Winnewisser, G.: J. Mol. Struct. (THEOCHEM) **742**, 215 (2005)
Owens, A., Yurchenko, S.N., Yachmenev, A., Thiel, W., Tennyson, J.: MNRAS **471**, 5025 (2017a)
Owens, A., Zak, E.J., Chubb, K.L., Yurchenko, S.N., Tennyson, J., Yachmenev, A.: Sci. Rep. **7**, 45068 (2017b)
Pachucki, K., Komasa, J.: Phys. Rev. A **78**, 052503 (2008)
Patrascu, A.T., Tennyson, J., Yurchenko, S.N.: MNRAS **449**, 3613 (2015)
Paulose, G., Barton, E.J., Yurchenko, S.N., Tennyson, J.: MNRAS **454**, 1931 (2015)
Pavlyuchko, A.I., Yurchenko, S.N., Tennyson, J.: MNRAS **452**, 1702 (2015)
Pickett, H.M., Poynter, R.L., Cohen, E.A., Delitsky, M.L., Pearson, J.C., Müller, H.S.P.: Quant. Spectrosc. Radiat. Transf. **60**, 883 (1998)
Polyansky, O.L., Zobov, N.F., Viti, S., Tennyson, J., Bernath, P.F., Wallace, L.: J. Mol. Spectrosc. **186**, 422 (1997)
Polyansky, O.L., Kyuberis, A.A., Lodi, L., Tennyson, J., Ovsyannikov, R.I., Zobov, N.: MNRAS **466**, 1363 (2017)
Pont, F., Sing, D.K., Gibson, N.P., Aigrain, S., Henry, G., Husnoo, N.: MNRAS **432**, 2917 (2013)
Prajapat, L., Jagoda, P., Lodi, L., Gorman, M.N., Yurchenko, S.N., Tennyson, J.: MNRAS **472**, 3648 (2017)
Ram, R.S., Brooke, J.S.A., Western, C.M., Bernath, P.F.: J. Quant. Spectrosc. Radiat. Transf. **138**, 107 (2014)
Rey, M., Nikitin, A.V., Babikov, Y.L., Tyuterev, V.G.: J. Mol. Struct. (Theochem) **327**, 138 (2016)
Rey, M., Nikitin, A.V., Tyuterev, V.G.: ApJ **847**, 105 (2017)
Rivlin, T., Lodi, L., Yurchenko, S.N., Tennyson, J., Le Roy, R.J.: MNRAS **451**, 5153 (2015)
Rothman, L.S., Gordon, I.E., Barber, R.J., et al.: J. Quant. Spectrosc. Radiat. Transf. **111**, 2139 (2010)
Ryabchikova, T., Piskunov, N., Kurucz, R.L., Stempels, H.C., Heiter, U., Pakhomov, Y., Barklem, P.S.: Phys. Scr. **90**, 054005 (2015)
Sauval, A.J., Tatum, J.B.: ApJS **56**, 193 (1984)
Semenov, M., Yurchenko, S.N., Tennyson, J.: J. Mol. Spectrosc. **330**, 57 (2017)
Sharpe, S.W., Johnson, T.J., Sams, R.L., Chu, P.M., Rhoderick, G.C., Johnson, P.A.: Appl. Spectrosc. **58**, 1452 (2004)

Sochi, T., Tennyson, J.: MNRAS **405**, 2345 (2010)
Sousa-Silva, C., Yurchenko, S.N., Tennyson, J.: J. Mol. Spectrosc. **288**, 28 (2013)
Sousa-Silva, C., Al-Refaie, A.F., Tennyson, J., Yurchenko, S.N.: MNRAS **446**, 2337 (2015)
Tennyson, J.: Astronomical Spectroscopy: An Introduction to the Atomic and Molecular Physics of Astronomical Spectra, 2nd edn. World Scientific, Singapore (2011)
Tennyson, J., Yurchenko, S.N.: MNRAS **425**, 21 (2012)
Tennyson, J., Bernath, P.F., Campargue, A., et al.: Pure. Appl. Chem. **86**, 1931 (2014)
Tennyson, J., Hulme, K., Naim, O.K., Yurchenko, S.N.: J. Phys. B: At. Mol. Opt. Phys. **49**, 044002 (2016a)
Tennyson, J., Lodi, L., McKemmish, L.K., Yurchenko, S.N.: J. Phys. B: At. Mol. Opt. Phys. **49**, 102001 (2016b)
Tennyson, J., Yurchenko, S.N., Al-Refaie, A.F., et al.: J. Mol. Spectrosc. **327**, 73 (2016c)
Tinetti, G., Vidal-Madjar, A., Liang, M.C., et al.: Nature **448**, 169 (2007)
Tinetti, G., Tennyson, J., Griffiths, C.A., Waldmann, I.: Phil. Trans. R. Soc. Lond. A **370**, 2749 (2012)
Townes, C.H., Schawlow, A.L.: Microwave Spectroscopy. Dover (2012)
Trafton, L.M., Miller, S., Geballe, T.R., Tennyson, J., Ballester, G.E.: ApJ **524**, 1059 (1999)
Underwood, D.S., Tennyson, J., Yurchenko, S.N., Clausen, S., Fateev, A.: MNRAS **462**, 4300 (2016a)
Underwood, D.S., Tennyson, J., Yurchenko, S.N., Huang, X., Schwenke, D.W., Lee, T.J., Clausen, S., Fateev, A.: MNRAS **459**, 3890 (2016b)
Vidal-Madjar, A., des Etangs, A.L., Desert, J.M., Ballester, G.E., Ferlet, R., Hebrand, G., Mayor, M.: Nature **422**, 143 (2003)
Vidler, M., Tennyson, J.: J. Chem. Phys. **113**, 9766 (2000)
Voronin, B.A., Tennyson, J., Tolchenov, R.N., Lugovskoy, A.A., Yurchenko, S.N.: MNRAS **402**, 492 (2010)
Wallace, L., Livingston, W., Hinkle, K., Bernath, P.: ApJS **106**, 165 (1996)
Wende, S., Reiners, A., Seifahrt, A., Bernath, P.F.: A&A **523**, A58 (2010)
Wenger, C., Champion, J.P.: J. Quant. Spectrosc. Radiat. Transf. **59**, 471 (1998)
Wong, A., Yurchenko, S.N., Bernath, P., Mueller, H.S.P., McConkey, S., Tennyson, J.: MNRAS **470**, 882 (2017)
Woodgate, G.K.: Elementary Atomic Structure, 2nd edn. Oxford University Press (1983)
Yadin, B., Vaness, T., Conti, P., Hill, C., Yurchenko, S.N., Tennyson, J.: MNRAS **425**, 34 (2012)
Yorke, L., Yurchenko, S.N., Lodi, L., Tennyson, J.: MNRAS **445**, 1383 (2014)
Yurchenko, S.N.: J. Quant. Spectrosc. Radiat. Transf. **152**, 28 (2015)
Yurchenko, S.N., Tennyson, J.: MNRAS **440**, 1649 (2014)
Yurchenko, S.N., Barber, R.J., Tennyson, J.: MNRAS **413**, 1828 (2011)
Yurchenko, S.N., Tennyson, J., Bailey, J., Hollis, M.D.J., Tinetti, G.: Proc. Nat. Acad. Sci. **111**, 9379 (2014)
Yurchenko, S.N., Blissett, A., Asari, U., Vasilios, M., Hill, C., Tennyson, J.: MNRAS **456**, 4524 (2016a)
Yurchenko, S.N., Lodi, L., Tennyson, J., Stolyarov, A.V.: Comput. Phys. Commun. **202**, 262 (2016b)
Yurchenko, S.N., Amundsen, D.S., Tennyson, J., Waldmann, I.P.: A&A **605**, A95 (2017)
Yurchenko, S.N., Sinden, F., Lodi, L., Hill, C., Gorman M.N., Tennyson, J.: MNRAS **473**, 5324 (2018a)
Yurchenko, S.N., Al-Refaie, A.F., Tennyson, J.: A&A (2018b)

Part IV
Solar System Atmospheres

Chapter 4
Atmospheric Physics and Atmospheres of Solar-System Bodies

Davide Grassi

Abstract The physical principles governing the planetary atmospheres are briefly introduced in the first part of this chapter, moving from the examples of Solar System bodies. Namely, the concepts of collisional regime, balance equations, hydrostatic equilibrium and energy transport are outlined. Further discussion is also provided on the main drivers governing the origin and evolution of atmospheres as well as on chemical and physical changes occurring in these systems, such as photochemistry, aerosol condensation and diffusion. In the second part, an overview about the Solar System atmospheres is provided, mostly focussing on tropospheres. Namely, phenomena related to aerosol occurrence, global circulation, meteorology and thermal structure are described for rocky planets (Venus and Mars), gaseous and icy giants and the smaller icy bodies of the outer Solar System.

4.1 Introduction

4.1.1 Definitions

The term 'atmosphere' is used to indicate the outermost gaseous parts of a celestial body (planets, moons and minor bodies, in the case of our own Solar System) and retained bounded by its gravity. As such, the term encompasses a large variety of structures, ranging from the massive envelopes that form the visible parts of giant planets, to the near-vacuum conditions found close to the surfaces of Mercury and of the Moon.

From a physical perspective, several aspects shall be considered for characterizing the conditions of an atmosphere. The list below describes, qualitatively, the most relevant ones:

D. Grassi (✉)
INAF – Institute for Space Astrophysics and Planetology, Via Fosso del Cavaliere 100, 00133 Rome, Italy
e-mail: davide.grassi@iaps.inaf.it

V. Bozza et al. (eds.), *Astrophysics of Exoplanetary Atmospheres*, Astrophysics and Space Science Library 450, https://doi.org/10.1007/978-3-319-89701-1_4

- *Collisional regime*: the collisions between the different atoms of a gas ensure the distribution of energy and momentum between the individual components. When density becomes so low that individual atoms and molecules simply behave as quasi-collisionless ballistic projectiles, we are dealing with an *exosphere*. In these conditions, the dynamic of atmospheric components can no longer be described with usual hydrodynamic equations and a more complex treatment is required.
- *Ionization*: the individual atoms/molecules composing an atmosphere may become ionized by the action of UV solar radiation or—in lesser extent—by the impinging of energetic particles of external origin. In conditions of low absolute density, the recombination of ionization products is slower than the production rate and ions may accumulate. In this case we have a *ionosphere*, whose ionic components are affected by magnetic fields, either generated from the parent body or of external origin (such as the magnetic field transported by the solar wind).
- *Local thermodynamical equilibrium*: statistical mechanics predicts that a distribution of molecules, over the possible quantum roto-vibrational states for a given species, follows the Boltzmann distribution, and is therefore driven exclusively by the gas temperature. When particles become more rarefied, collisions are less effective in distributing energy by collisions among individual particles and other factors (namely, absorption of Solar infrared and visible radiation) may create substantial deviations in energy distribution. The on-set of non-LTE conditions has not an immediate impact on dynamical behaviour of gases, but may influence substantially the overall energy balance of the atmosphere.
- *Turbulent regime*: the turbulence in the lowest parts of the atmosphere is ensured by mixing a uniform composition along altitude (*homosphere*). Above the turbopause, diffusion phenomena becomes dominant and substantial fractionation of species, due to different molecular weight, may occur.
- *Energy transport*: planetary atmospheres can be seen as systems that perform a net transport of energy from their lowest parts toward space, where energy is eventually dispersed mostly as electromagnetic radiation. Ultimate sources of energy in the deep atmospheres are the absorption of solar radiation (notably, by solid surfaces) or heat from deep interiors, accumulated either by conversion of kinetic energy (dissipated during accretion or by gravitational differentiation) or by decay of radioactive elements. In the lowest parts of the atmospheres energy is usually transported by convection, while in the upper parts radiative processes become dominant.
- *Absorption of solar radiation*: it is particularly important in the ultraviolet and in the Near Infrared spectral range, where a number of common atmospheric components are optically active and the Sun spectrum retains a significant flux. Moving downward from the top of the atmosphere, the progressive increase of atmospheric densities makes the overall heating, by absorption of Solar Flux, effective in rising the air temperature above the values expected by simple transport from deeper levels. However, moving further downward, a progressively lesser amount of radiation reaches the deepest levels and the energy deposition becomes less and less effective. The region below the peak efficiency of the process is often characterized by a decrease of temperature moving downward and is called

stratosphere. The name is justified by the limited vertical motions observed in this regions, inhibited by the gravitationally stable temperature structure (cold dense layers below, warm lighter layers above). Albeit rather common, the existence of a stratosphere requires the presence of gases optically active in the UV and is therefore absent on Mars and Venus.
- *Aerosol condensation*: the decrease of air temperatures with altitude, typical of the lowest parts of the atmosphere, determines the ubiquitous occurrence of aerosols observed in the Solar System atmospheres. The aerosols makes atmospheric motions immediately observable in the so-called *troposphere*. If a stratosphere exists above, the local air temperature minimum is usually designated as *tropopause*.

4.1.2 Collisional and Non-collisional Regimes

Most of the studies on atmosphere physics make an implicit assumption on the validity of fluid dynamics. In the perspective of the extension of these concepts to the atmospheres of exoplanets - where the most exotic conditions can not be ruled out a priori - it is useful to consider the basis of this assumption. Fluid dynamics is derived from the continuum hypothesis, which states that mass of the fluid is distributed continuously in the space. This hypothesis essentially requires that the phenomena of our interest occurs at spatial scales greater than those related to the spacings of individual components (molecules or atoms) of the fluid. Since momentum exchange between particles takes essentially the form of collisions, it is convenient to introduce the so-called Knudsen number, defined as the ratio between the mean free path of gas molecules λ and the typical scale length L of the considered phenomena:

$$K_{\mathrm{n}} = \frac{\lambda}{L} \, . \tag{4.1}$$

When $K_{\mathrm{n}} << 1$, collisions are so frequent that continuum hypothesis is satisfied and fluid dynamics can be applied. When K_{n} is in the order of 1 or above, the collisions among particles are no longer enough frequent to ensure the validity of the continuum assumption and a more general approach is required.

In order to describe, in the most general terms, the behaviour of a gas, let us consider—for simplicity—a system formed by N identical particles. Considering $dx\,dv$ as the infinitesimal element in the phase space (where coordinates are the spatial positions and the speeds along the three dimensions), we can define a distribution function f such that

$$\int_{R^3}\int_{R^3} f\,(x, v; t)\,dx\,dv = N \, . \tag{4.2}$$

The behaviour of the system is described by the temporal evolution of the distribution function f. The Boltzmann equation states that

$$\frac{df}{dt} = \left(\frac{\partial f}{\partial t}\right) + \vec{v} \cdot \vec{\nabla}_x f = \left(\frac{\partial f}{\partial t}\right)_{\text{force}} + \left(\frac{\partial f}{\partial t}\right)_{\text{collision}} . \tag{4.3}$$

In the first passage, we made explicit the diffusion coefficient, i.e. variations of distribution functions due to the undisturbed motion of particles. The two terms on the right-hand side describe the results of the external forces acting on the particles of the system (such as gravity) and the collisions occurring among the particles of the system, respectively.

This leads to

$$\left(\frac{\partial f}{\partial t}\right) = -\vec{v} \cdot \vec{\nabla}_x f - \frac{1}{m}\vec{F} \cdot \vec{\nabla}_v f + \left(\frac{\partial f}{\partial t}\right)_{\text{collision}} , \tag{4.4}$$

being m the mass of individual particles and $\vec{F}$ the force acting on individual particles.

The collisional term can be expressed as:

$$\left(\frac{\partial f}{\partial t}\right)_{\text{collision}} = \int_{R^3}\int_{R^3}\int_{S^2} B(g, \Omega)\left[f(v'_{\text{A}}, t)\, f(v'_{\text{B}}, t) - f(v_{\text{A}}, t)\, f(v_{\text{B}}, t)\right] d\omega\, dv_{\text{A}}\, dv_{\text{B}} , \tag{4.5}$$

where the apex (or its absence) indicates the particle after (or before) the collision,

$$g \equiv |v_{\text{A}} - v_{\text{B}}| = |v'_{\text{A}} - v'_{\text{B}}| \tag{4.6}$$

is the magnitude of relative speed, Ω indicates the angular change in relative speeds after collision, and $B(g, \Omega)$ is a collision kernel providing the cross section of the collision. A detailed introduction on these subjects is provided by Pareschi (2009).

Albeit simplifications have been proposed for the modelling of the collisional term, in matter of fact numerical methods—such as the Direct Simulation Monte Carlo (DSMC: Bird 1970)—are usually adopted for the treatment of rarefied gases. Namely, the DSMC method has been applied successfully to the modelling of planetary exospheres (e.g. Shematovich et al. 2005).

4.1.3 Balance Equations for Mass, Momentum and Energy

When $K_{\text{n}} << 1$, the fluid behaviour can be modelled according the principles of fluid dynamics. In this approach, the fluid is considered as composed of a set of deformable volumes and properties such as density and temperature are considered as continuous fields, defined at infinitesimal scale, completely neglecting the actual molecular nature of the fluid. In considering the behaviour of the fluid, two possible approaches can be adopted. In the Eulerian perspective, the volumes are defined by their position in a fixed (not time-variable) spatial reference frame, and the fluid is observed while flowing across this ideal grid. In the Lagrangian perspective, the individual fluid volumes retain their identity while moving in the space and possibly being deformed during the motion. The two approaches are related through the definition of the material derivative. For any given scalar field a being a function of spatial coordinates $\vec{x}$ and time t, and being $\vec{u}$ the fluid speed, the material derivative (i.e. the time derivative as seen by a Lagrangian observer) is given by:

$$\frac{Da}{Dt} = \frac{\partial a}{\partial t} + \vec{u} \cdot \vec{\nabla} a\,. \tag{4.7}$$

The Lagrangian approach is often adopted in introducing the principles of fluid dynamics since they can be derived from the general principles of conservation of momentum and mass as applied to the infinitesimal volumes.

Being ρ the density, the conservation of mass is described by

$$\frac{D\rho}{Dt} + \rho\, \vec{\nabla} \cdot \vec{u} = 0\,. \tag{4.8}$$

This equation simply states that variations of density are related to net flow of mass to/from the reference volume.

The general form of momentum balance (known as the Naiver-Stokes equation) can be considerably simplified for the gas case as follows

$$\rho \frac{D\vec{u}}{Dt} = \rho\, \vec{g} - \vec{\nabla} P + \vec{\nabla} \cdot \left(\mu\, \nabla^2 \vec{u}\right), \tag{4.9}$$

where P is the pressure, $\vec{g}$ the acceleration induced by gravity (and any other force field acting on the entire fluid) and μ is the coefficient of viscosity of the fluid. The equation states that variation of momentum for the reference volumes can be induced by an external force, a net gradient of pressure and frictional drag.

The energy balance is derived directly from the first law of thermodynamic and is expressed by

$$\rho\, c_{\mathrm{p}} \frac{DT}{Dt} = -P\, \vec{\nabla} \cdot \vec{u} - \vec{\nabla} \cdot \vec{F} + k \nabla^2 T + \rho \dot{q} \tag{4.10}$$

where here $\vec{F}$ is now the radiative flux, T is the temperature, k the thermal conductivity, $\dot{q}$ the internal heating rate, and c_{p} the specific heat at constant pressure. The energy varies therefore because of the performed mechanical work, thermal diffusion, net radiative balance and internal sources of heating (notably, release/absorption of latent heat associated to phase changes).

Pareschi (2009) provides hints on the formal derivation of Eqs. (4.8)–(4.10) as limit case of the Boltzmann equation. An introduction more focused on atmospheric dynamic is given by Salby (1996, Chap. 10).

4.1.4 Turbulence

The balance equations described in the previous section allow one to describe the motion of air masses in a large variety of conditions. A major complication is represented by the onset of *turbulence*. This conditions holds when the relative motion of fluid particles can no longer be represented as a laminar flow, where parcels moves along quasi-parellel layers with limited relative mixing. The turbulent flow is characterized by the on set of *eddies* (areas where a fluid tends to rotate around a preferential axis) at different spatial scales, that allows the energy and momentum related to the fluid motion to be effectively distributed also along the directions orthogonal to the original motion. Another typical behaviour of the turbulent motion is represented by the chaotic variations of fields such as pressure and air speed both in time as well as along the spatial coordinates. Albeit air parcels with sizes smaller than typical eddies still follows the balance equations, it becomes impossible to predict exactly the detailed behaviour of the overall system. A useful parameter to describe the behaviour of a fluid with respect to the turbulence is the Reynolds number

$$R_{\mathrm{n}} = \frac{\rho\, u\, L}{\mu}, \tag{4.11}$$

being L a characteristic length of the considered phenomenon. For $R_{\mathrm{n}} < 2000$ the motions are typically laminar since the viscous shear can effectively distribute the energy among contiguous layers. For $R_{\mathrm{n}} > 5000$ the motions are typically turbulent and eddies tend to develop.

Albeit a complete mathematical treatment of turbulence is still missing, existing theory allows to infer some key properties. Eddies in turbulence are organized along different spatial scale, with a cascade transfer of energy toward smaller structures. At scales below the *Kolmogorov scale length*, the viscous dissipation eventually convert the kinetic energy into heat. In the case of atmospheres, the eddies related to turbulence may reach sizes of several kilometres up to hundreds of kilometres in the giant planets. Conversely, the Kolmogorv scale length is several orders of magnitude smaller, typically on millimetre scale.

4.1.5 Overall Structure of the Atmosphere

The simplest possible treatment of the structure of the planetary atmospheres starts from the assumption of stationary conditions, with no atmospheric motions.

In this case, only considering the vertical direction along z, the momentum balance, Eq. (4.9), states that increments in pressure are due to variations in the overlying atmospheric column

$$dP = n(z)\, g(z)\, \mu_{\mathrm{a}}(z)\, u\, dz\,, \tag{4.12}$$

being n the molecular number density, μ_{a} the mean molecular weight, g the gravity acceleration (all these quantities being a function of altitude z) and u the unified atomic mass unit. This condition, called *hydrostatic equilibrium*, is usually assumed to hold in the Solar System atmospheres and interiors. Pressure deviations associated to the actual vertical motions are typically extremely small and even in the case of motions involving large masses of air (such as Hadley circulation, see Sect. 3.2), hydrostatic equilibrium can be safely assumed.

The perfect gas law

$$P = n\, k_{\mathrm{b}}\, T\,, \tag{4.13}$$

where k_{b} is the Boltzmann constant, holds in a large range of conditions found in the atmospheres of Solar System for low Knudsen numbers. Its differentiation leads to

$$dn = -n(z)\left(\frac{1}{T(z)}\frac{dT}{dz} + \frac{\mu_{\mathrm{a}}(z)\, g(z)\, u}{k_{\mathrm{b}}\, T(z)}\right) dz \equiv -n(z)\frac{1}{H(z)}\, dz\,. \tag{4.14}$$

This equation defines the atmospheric scale height H, as the quantity that locally governs the variation of density with altitude. Once one considers that both μ_{a} and g are rather slow functions of altitude, it becomes evident how the overall density structure of atmospheres is essentially driven by its temperature structure.

In assessing the overall energy budget of an atmosphere, several factors must be taken into account.

Inputs:

- Direct absorption of incoming radiation. Absorption of UV solar radiation is the main responsible for heating in the upper atmospheres of planet, being usually associated to the occurrence of stratosphere. Absorption of infrared solar photons—albeit associated to intrinsically low fluxes—can be important in the most opaques spectral regions in centres of main bands of IR active species.
- Solar heating of the surface: solar visible radiation is effectively absorbed by planetary surfaces, that are therefore an indirect source of heating for overlying atmospheres.
- Heat from interior: is the main source of energy for the atmospheres of giant planets. It is caused by the still ongoing cooling of the interior from the heat accumulated during the accretion phase (kinetic energy of impactors was converted into heat). A secondary source is represented, for Jupiter and possibly Saturn, by the precipitation of helium and argon toward the centre through the metallic hydrogen mantle.

Other mechanisms may become important in the upper atmospheric layers, like:

- Precipitation of charged energetic particles: planets with substantial intrinsic magnetic fields are subject to precipitation of charged particles having been accelerated in the magnetosphere. Precipitation is often made evident by the occurrence of auroras.
- Joule heating: associated to the electric current systems developing in the ionospheres of planets with an intrinsic magnetic field.

Outputs:

- Emission of infrared radiation: given the typical temperatures found in the Solar-System atmospheres, the corresponding thermal emissions peak in the infrared domain. Infrared emission is by far the most important mechanism of net loss of energy from the atmospheres and is governed by the presence of IR-active molecules, most important ones being methane, carbon dioxide and water.

Transport between different parts of the atmosphere:

- Radiative transfer: it consists in the net transfer of energy from warm layers to colder ones by means of IR photons. Efficiency of transfer is inhibited by high total opacities between involved layers. Absorption is important in most dense parts of the atmospheres, given the higher number of active molecules per unit of optical path length. Here, in presence of IR active species, radiation thermally emitted by the surface (upon absorption of Solar visible radiation) or by lower atmospheric layers is promptly absorbed and represent the basis of the so-called *greenhouse effect.*
- Convection: air parcels warmed in the deepest parts of atmospheres become buoyant with respect to the surrounding environment and move upward. Their vertical motion represent an efficient mechanism for vertical transport of energy and minor atmospheric components. Horizontal displacements of air masses related to global circulation (see Sect. 3.2, ultimately driven by convection) are the main transport mechanism of energy between different latitudes.
- Conduction: is the main exchange mechanism between atmosphere and surface. In other parts of the atmosphere is usually less important than radiative transfer and convection, but becomes again important in the *thermosphere*.
- Phase changes: latent heat of vaporization/condensation associated to aerosols may become a dominant term in energy budget of tropospheres. A particularly important effect is the enhancement of convection associated to cloud formation.

4.1.6 Equilibrium Temperature of Planetary Surfaces

Solar radiation is the main source of energy driving the phenomena occurring in the atmospheres of rocky planets, satellites and minor bodies of the solar system. In order to discuss these aspects in more detail, some further nomenclature shall be introduced. An excellent introduction to the role of radiation in planetary atmospheres can be found in Hanel et al. (2003).

The radiation intensity I_ν is defined considering the amount of energy transported by radiation propagating at angle θ over a surface of area ds in time dt within the solid angle $d\omega$ and the frequency interval $d\nu$

$$I_\nu = \frac{dE}{dt\, d\omega\, ds\, \cos\theta\, d\nu} . \tag{4.15}$$

A particularly important case of radiation intensity is the one describing the thermal emission by a *black body*, an ideal system capable to adsorb completely any incoming radiation.

$$B_\nu = \frac{2\, h\, \nu^3}{c^2} \frac{1}{e^{\frac{h\,\nu}{k_B\, T}} - 1} , \tag{4.16}$$

where h is the Planck constant and c the light speed.

The integration of Eq. (4.16) over frequencies leads to

$$\int_0^\infty B_\nu\, d\nu = \frac{2\,\pi^5\, k_B^4}{15\, h^3\, c^2} T^4 \stackrel{\text{def}}{=} \sigma\, T^4 . \tag{4.17}$$

where σ is referred as the Stefan-Boltzmann constant. The actual thermal radiation emitted from a planetary surface is often described in terms of emissivity ϵ, defined such as

$$I_\nu = \epsilon_\nu\, B_\nu(T) . \tag{4.18}$$

The radiative net flux F_ν along a given direction is defined considering the direction as $\theta = 0$ and then integrating intensity over all solid angles

$$F_\nu = \int I_\nu \cos\theta\, d\omega . \tag{4.19}$$

This quantity represent the net amount of energy transported by radiation over the surface ds orthogonal to the direction $\theta = 0$. The total radiative flux F is just the integration of F_ν over the entire spectrum.

A first rough estimate of surface equilibrium temperatures T_{eq} for bodies with a solid surface can be performed equating the total flux absorbed from Solar radiation to thermally-emitted infrared radiation. Neglecting the temperature variations induced by planetary rotation and latitude, this requirement becomes

$$\frac{F}{a^2}(1 - A)\,\pi\, r^2 = \sigma\, \epsilon\, T_{\text{eq}}^4\, 4\,\pi\, r^2 \Rightarrow T_{\text{eq}} = \left(\frac{F}{a^2}\frac{(1 - A)}{4\,\sigma\,\epsilon}\right)^{\frac{1}{4}} , \tag{4.20}$$

where F is the total solar flux at 1 astronomical unit (au), a is the planet-Sun distance, A and ϵ the spectrally averaged albedo (reflectance) and emissivity, respectively.

Table 4.1 compares the expected and observed surface temperatures of the Solar-System rocky planets, the Moon and Titan. It is evident how the Earth and Venus present surface temperatures much higher than the expected ones. This is mostly due

Table 4.1 Comparison between the expected and observed surface temperatures of the Solar-System rocky planets, the Moon and Titan

	A	a (au)	T_{eq} (K)	T_{surf} (K)
Mercury	0.11	0.387	440	100–700
Venus	0.72	0.723	230	740
Moon	0.07	1	270	100–400
Earth	0.36	1	256	290
Mars	0.25	1.52	218	223
Titan	0.22	9.6	85	93

to the effective trapping of energy due to the absorption by the atmosphere of the radiation thermally emitted by the surface.

4.1.7 Mechanisms for Energy Transfer in the Atmospheres

Convection is a fundamental phenomenon occurring in planetary atmospheres when an air parcel adsorbs energy at its lower boundary and, upon expansion, becomes less dense and buoyant with respect to the surrounding environment. The consequent vertical rise can, in a large range of conditions, be considered as an adiabatic process, where energy is exchanged with the surrounding environment only in the form of mechanical work.

In the assumption of a negligible role from latent heat (dry convection), we can infer the expected vertical temperature profile for a convective layer of the atmosphere. Let us consider the first law of thermodynamic

$$dQ = dU + P\,dV\,, \tag{4.21}$$

where dQ is the heat exchange with the environment, $P\,dV$ the mechanical work and dU the variation of internal energy. By definition we have an adiabatic process whenever $dQ = 0$. Taking into account the definitions of the of specific heat at fixed volume and pressure C_V and C_P, we can demonstrate that the adiabatic conditions implies

$$C_P\,dT = \frac{dP}{\rho}\,. \tag{4.22}$$

Considering furthermore the condition of hydrostatic equilibrium, Eq. (12), we infer

$$\frac{dT}{dz} = -\frac{g}{C_P}\,. \tag{4.23}$$

On the Earth atmosphere, the (dry) adiabatic lapse rate is about $9.5\,\mathrm{K\,km^{-1}}$. The adiabatic lapse rate represent a maximum limit to the vertical temperature gradient. Every local increase of vertical gradient above this limit usually prompt a quick onset of convection, that allows an efficient way to transport excess heat in higher parts of the atmosphere. Probe observations demonstrated how temperature profiles in the deep ($P >> 1$ bar) atmospheres of Venus and Jupiter lie very close to local values of adiabatic lapse rate.

The other fundamental mechanism for energy transport in planetary atmospheres is given by radiation. An atmosphere is said to be in radiative equilibrium when

$$\frac{dF}{dz} = 0\,. \tag{4.24}$$

To derive the corresponding atmospheric temperature profile, we will assume that radiative equilibrium holds for every frequency and that atmosphere is optically thick, implying that locally the radiation intensity follows the Planck distribution. Under these conditions it can be demonstrated that

$$F(z) \propto \frac{1}{\rho}\frac{\partial T}{\partial z}\int_0^\infty \frac{1}{\alpha_\nu}\frac{\partial B_\nu}{\partial T}d\nu\,, \tag{4.25}$$

where α_ν is the extinction coefficient per mass unit. Considering now a mean extinction coefficient α_{mean}, we have

$$\frac{dT}{dz} \propto \frac{\alpha_{\mathrm{mean}}\,\rho\,F}{\sigma\,T^3}\,. \tag{4.26}$$

The thermal gradient becomes therefore steeper with increasing mean opacity, which makes the transport of energy through the atmosphere less and less effective. In planetary atmospheres, transfer by radiative mechanisms occurs typically at the intermediate opacity regimes seen in stratospheres: at lower levels, the higher infrared opacity makes convection more efficient (i.e.: has a lower lapse rate); at higher levels, low density allows greater free paths for molecules and direct thermal conduction becomes more effective due to effectively high thermal conductivity.

Details of these derivations can be found, together with an extensive introduction to planetary atmospheres, in de Pater and Lissauer (2010).

4.1.8 *Typical Temperature Profiles for the Atmospheres of Solar-System Planets*

Despite the large temporal and spatial variabilities on different scales observed within individual planetary atmospheres, it is possible to determine some significant mean temperatures profiles. Figure 4.1 presents typical cases from thick Solar-System atmospheres.

As discussed above, the deepest parts of atmospheres are dominated by the infrared opacities, that makes the radiation transport poorly effective. In the case of giant planets, main source of infrared opacity is represented by the collision-induced absorption (CIA) of molecular hydrogen. For other cases, infrared opacity is largely dominated by a minor constituent of the atmosphere (methane for Titan, carbon dioxide for the Earth), that turns out therefore to have a major impact on the overall thermal structure of the planet. At the approximate pressure level of 1 bar, mean IR opacity of the Solar-System atmospheres is still in the range between 2 and 9. Only at lower pressures, when opacity becomes in the order of unity, radiation can be effectively emitted toward deep space. This is the approximate level where the boundary between convective and radiative region is located.

At higher altitudes, opacity at shorter wavelengths (UV) become dominant over IR opacity. The latter is indeed critically dependent upon pressure, given the dependence of CIA and line broadening upon pressure. Opacity at shorter wavelengths is due to a number of minor components (notably methane and ozone) that experience photo dissociation, with a net deposition of energy from the Sun directly into the atmosphere, with an increase in air temperature that creates the observed stratosphere (as defined by the positive lapse rate). Robinson and Catling (2014) demonstrated that in the case of upper atmospheres with a short-wave optical depth much greater than infrared optical depth, the stratopause develops - for a large range of conditions - at the approximate level of 0.1 bar. The same study indicates that stratopause develops always well inside the radiative region of the atmosphere.

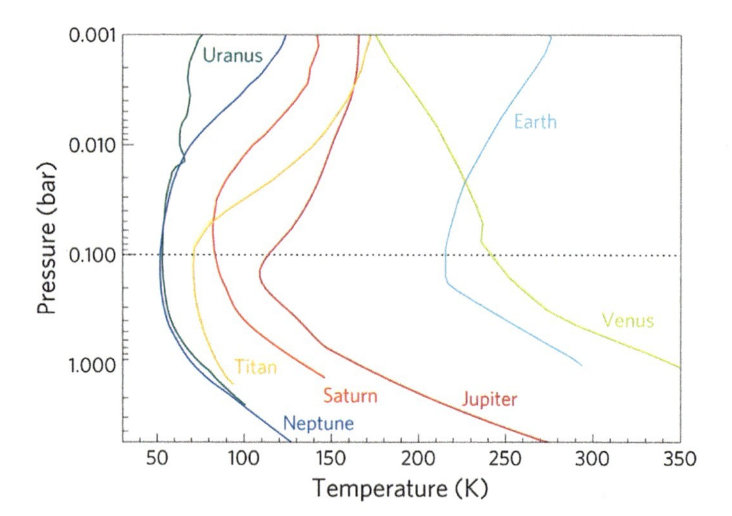

Fig. 4.1 Mean air temperature versus pressure profiles for several Solar-System atmospheres. Reprinted by permission from Macmillan Publishers Ltd: Nature, Robinson and Catling (2014), Copyright 2013

4.2 Physical and Chemical Changes in Planetary Atmospheres

4.2.1 *Origin of Planetary Atmospheres*

Table 4.2 summarizes the mean composition of planetary atmospheres of the Solar System, at the respective surfaces or at the approximate 1 bar level for the giant planets. On the basis of the observed composition, we can distinguish two main cases:

- Giant planets are believed to largely retain in their gaseous envelopes the original elemental compositions of the planetoids they form from. However, tropospheric levels reachable by remote sensing techniques or by probes can present substantial depletions with respect to an ideal 'well mixed' composition due to condensation of aerosols at deeper levels, local meteorology or phase changes capable to induce fractionation of specific elements (most notable example being the insolubility of helium in metallic hydrogen).
- Atmospheres of rocky planets are believed to be largely "secondary", i.e.: formed by out gassing of planetary interior after that the original gaseous envelope, accreted during the planet formation, was removed by the action of solar radiation and wind in the early life of the Sun or by impacts. While the weak gravity fields of rocky planets could possibly justify the gravitational escape of lighter elements such as Hydrogen and Helium, this would leave behind substantial (albeit not intact) amount of heavier noble gases, in amounts well above the trace levels currently observed. Later contributions to the secondary atmospheres from impacts (notably, comets) may also have been important. Given the large variety of present-day conditions observed in rocky planets, subsequent evolution—directly or indirectly driven by the distance from the Sun—has also been fundamental.

For giant planets, the comparison of elemental abundances observed in atmosphere against the ones inferred for the Sun may provide key constraints on the formation scenarios (Fig. 4.2). Best experimental evidence is currently available for Jupiter, where a direct sample was performed by the *Galileo Entry Probe* during its descent on Dec 7th, 1995. Overall, heavy elements appear to be

Table 4.2 Mean troposphere composition (volume percentages) and surface pressure of atmospheres of the Solar System. Updated from NASA Planetary Fact Sheets (https://nssdc.gsfc.nasa.gov/planetary/factsheet/)

Venus	Earth	Mars	Jupiter (%)	Saturn (%)	Uranus (%)	Neptune (%)	Titan	Pluto (and Triton)
CO_2 96.5%	N_2 78%	CO_2 96.0%	H_2 86.0	H_2 87.0	H_2 82.0	H_2 80.0	N_2 95%	N_2 95%
N_2 3.5%	O_2 21%	Ar 1.9%	He 13.0	He 12.0	He 15.0	He 19.0	CH_4 5%	CH_4 traces
–	–	N_2 1.9%	CH_4 0.3	CH_4 0.3	CH_4 2.5	CH_4 1.5	–	CO traces
90 bar	1 bar	0.06 bar	–	–	–	–	1.45 bar	10 μbar

enriched consistently by a factor 3 with respect to the mean solar composition, that would been contrarily expected in the case of a direct capture of material from the proto-solar nebula by direct gravitational accretion. The observed enrichment suggests therefore that a central core of the proto-Jupiter must have formed firstly. Only in subsequent phases, its gravitational attraction shall have been sufficient to attract planetoids rich in volatile elements (in forms of "ices") but lacking the hydrogen envelop due to their very low masses. Composition data are much more fragmentary for Saturn, where an enrichment of a factor 10 can be inferred from a limited number of infrared active molecules and for Uranus and Neptune, where an enrichment of about 30 is assumed on the solely basis of the estimate of carbon (in the form of methane). An extensive discussion on these aspects is provided by (Atreya et al. 2018). Ices are expected to have contributed in much larger extent to the formation of Uranus and Neptune ("icy giants") with respect to Jupiter and Saturn cases ("gas giants"). Indeed, albeit most external layers of icy giants are still formed mostly of Hydrogen and Helium, heavier elements are thought to represent a substantial fraction of planetary interiors, as demonstrated by their mean densities, with values of 1.27 and 1.63 g cm^3, similar to the Jupiter value (1.32 g cm^3), despite the much lower degree of internal compression.

For the cases of rocky planets, extensive volcanism is expected to have occurred at the beginning of the Solar-System existence, as a result of the dissipation of the internal heat, which was accumulated by impacts during the accretion phase and by decay of radioactive materials. Inference of the composition of the secondary atmospheres from the measurements on current volcanic gas releases on Earth volcanoes is however prone to substantial uncertainties. Most of volcanic activity on present-day Earth is associated to subsidence of oceanic crust and presents therefore

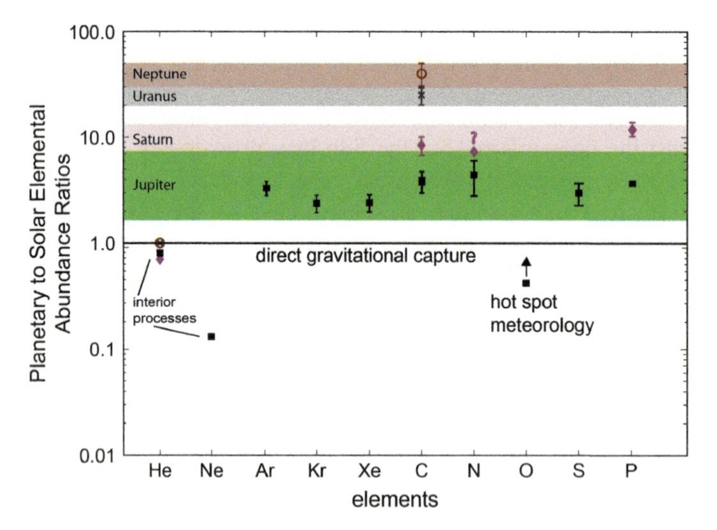

Fig. 4.2 Elemental enrichment with respect to the Solar composition in the atmospheres of giant planets. Courtesy PSL/ Univ. of Michigan (http://clasp-research.engin.umich.edu/psl/research.html). For an updated version, see Fig. 2.1 in Atreya et al. (2018)

substantial enhancement in water and carbon content associated to reprocessing of ocean floors. Possibly more representative are outgassing from "hot-spot" volcanoes away from plate boundaries, assuming that they originated at the core-mantle interface. In these cases, an enrichment in sulphur (a supposed component of the core) shall be expected. Despite large variability among different samples, hot-spot outgassing consist primarily of carbon dioxide (up to 50%), water (up to 40%) and sulphur dioxide (Symonds et al. 1994). Role and timing of post-formation impacts to the overall inventory of volatiles in the atmospheres of rocky planets is still matter of debate. The value observed for the hydrogen isotopic ratio D/H in the water of Earth oceans is about one third of that determined to exist in comet 67P (Altwegg et al. 2015), albeit substantial uncertainties exist on the overall significance of available cometary estimates with respect to the overall population of these objects. Conversely, the D/H ratio in carbon-rich chondroid meteorites matches quite well the Earth's values, pointing to a possible role of these reservoirs in forming the atmospheres of rocky planets (Morbidelli et al. 2000).

4.2.2 Loss Mechanisms for Planetary Atmospheres

A substantial fraction of the atmospheric mass of a planet can be removed during its evolution by several different mechanisms:

- *Jeans thermal escape*: in regions where Knudsen number approaches one, there is a tail in the Maxwellian distribution of air particles speeds that exceeds escape velocity. Corresponding air particles can be lost in space since further collisions are rare and they behave as ideal projectiles. The fraction of air particles with speed modules between v and $v + dv$ is given by

$$f(v)dv = const\ N \left(\frac{m}{k_B T}\right)^{\frac{3}{2}} v^2\, e^{-\frac{m\, v^2}{2 k_B T}}\, dv\,, \tag{4.27}$$

  where m is the mass of the molecule ($m = \mu_a u$). The efficiency of loss depends therefore both on atmosphere temperature as well as on air particle mass. This may eventually result in variations of isotopic ratios of a given species with respect to their original values (isotopic fractionation), the effect being more evident in lighter species.
- *Hydrodynamic escape*: it occurs when heavier atoms, which are not expected to be efficiently removed on the basis of the Jeans escape, are subject to a high number of collisions from escaping lighter atoms. Heavier species are therefore effectively dragged away from planet atmosphere by momentum transfer. The mechanism requires a very high rate of Jeans escape over lighter species, a condition not currently seen in Solar System atmospheres. It is expected to become important in the very hot atmospheres of exoplanets and having been significant in the earliest phases of rocky planets evolution.

- *Impact erosion*: the impact of large bodies (with sizes of the order of atmospheric scale height) may eject ballistically a seizable fraction of the atmosphere. Air molecules/atoms with speeds exceeding the escape velocity are lost in space. This mechanism is not expected to produce substantial isotopic fractionation.
- *Sputtering*: individual air particles may achieve speeds exceeding the escape velocity upon collision with energetic neutral particles (ENA) or ions originating outside the atmosphere.
- *Solar wind sweeping*: atmospheric particles previously ionized by UV radiation or by the impinging of energetic particles are trapped in the magnetic-field lines of the Solar wind moving in the vicinity of upper ionosphere. The mechanism is important for planets not protected by a significant intrinsic magnetic field, such as Venus and Mars.

The mechanisms listed above remove permanently the atmospheric mass from the planet. However, the atmosphere can also be substantially depleted by a variety of other mechanisms that fix—permanently on temporary—a specific component of the atmosphere to the surface of the planet. Most important are condensation (notable examples are the water on the Earth to form oceans or the seasonal cycle of condensation/sublimation of carbon dioxide at the Mars poles) and chemical fixation (such as formation of carbonate or iron oxide deposits).

4.2.3 Evolution of the Atmospheres of Rocky Planets

4.2.3.1 Evolution of the Earth Atmosphere

A key factor on the evolution of Earth climate has been represented by the occurrence of large bodies of liquid water at the surface. This is due to the suitable orbital position of the planet, since closer distance to the Sun would have implied higher atmospheric temperatures and would have let the water in the form of steam. The presence of water was functional in removing substantial amounts of carbon from the atmosphere of the early Earth. The solution of carbon dioxide in water and subsequent reaction with silicate rocks are key steps leading eventually to the trapping of carbon in seabeds in the form of carbonates. An example can be the following

$$
\begin{aligned}
&\mathrm{CO_2}(aq) + \mathrm{H_2O} \rightarrow \mathrm{H_2CO_3}(aq) \\
&\mathrm{CaSiO_3}(s) + 2\mathrm{H_2CO_3}(aq) \rightarrow \mathrm{Ca^{2+}}(aq) + 2\mathrm{HCO_3^-}(aq) + \mathrm{H_2SiO_3}(aq) \\
&2\mathrm{HCO_3^-}(aq) \rightarrow \mathrm{CO_3^{2-}}(aq) + \mathrm{H_2O} + \mathrm{CO_2}(aq) \\
&\mathrm{Ca^{2+}}(aq) + \mathrm{CO_3^{2-}}(aq) \rightarrow \mathrm{CaCO_3}(s)
\end{aligned}
\tag{4.28}
$$

Current estimates on carbon pools expect the total amount trapped as carbonates to exceed by a factor about 5×10^4 the present day atmospheric content (Falkowski 2000). If released in the atmosphere as carbon dioxide, this would create a CO_2 dominated atmosphere with a surface pressure exceeding the one observed in Venus.

Another key step in the evolution of Earth atmosphere was the release of large amounts of molecular oxygen in the atmosphere. This was strictly linked to development of early life forms that evolved biological cycles producing free-oxygen as discard product. An example is the current form of photosynthesis, presented here in a strongly simplified form, that is

$$6CO_2 + 6H_2O \rightarrow C_6H_{12}O_6 + 6O_2 \,. \tag{4.29}$$

The molecular oxygen was initially fixed in rocks, oxidizing exposed iron-bearing minerals and forming the so called “red-bed formations” found worldwide. Once the geological sinks became saturated, the molecular oxygen begun to accumulate in the atmosphere. Availability of molecular oxygen lead to the creation of the ozone layer and to the oxidation of residual amounts of methane still present in the Earth atmosphere.

4.2.3.2 Evolution of the Venus Atmosphere

Albeit it is reasonable to assume that evolution of Venus started from conditions rather similar to those on the Earth, a substantial difference has been represented by the proximity to the Sun. It is generally assumed that liquid water may have existed on the surface during the earliest phases of Venus history (until 2 Gy ago). Nevertheless, the capture of carbonates has not been so effective as on the Earth since solubility of CO_2 decreases with increasing liquid temperature. The persistence of carbon dioxide in the atmosphere would have progressively risen the temperature and make the ocean evaporation more effective. The process has a positive feedback, being water vapour a greenhouse gas as well. This eventually resulted in a progressively faster rise of temperatures until the oceans completely evaporated (“runaway greenhouse”). If kept in the atmosphere, water vapour is more easily dissociated by UV radiation in the uppermost tropospheric levels. Given the low mass of hydrogen atoms, they are easily lost to space due to Jeans escape (Ingersoll 1969). Moreover, given the lack of substantial magnetosphere at Venus, ionized hydrogen atoms are more easily swept away by the Solar wind. A substantial loss of water in the Venus atmosphere is demonstrated by the D/H ratio measured in the very small amounts of water still present in the atmosphere, being this value about 150 times higher than the one observed on the Earth. This observation is consistent with the preferential loss of light species associated to the Jeans escape. Moreover, direct measurements of ion loss from Venus demonstrated that still today H^+ and O^+ are lost in space in stoichiometric ratios corresponding to those of water (Fedorov et al. 2011).

4.2.3.3 Evolution of the Mars Atmosphere

Surface of Mars bears clear evidence of the occurrence of liquid water in the geological past, but actual size of possible large water bodies on the surface is still matter of debate. Albeit gamma ray spectrometry has revealed that substantial amounts of war ice must exist in form of permafrost ice beneath the Mars surface (Boynton et al. 2002), the current surface pressure of the planet does not allow to sustain the occurrence of liquid water; consequently, long term climate changes shall have occurred along Mars history. Estimates based on argon isotopes confirm that the planet atmosphere shall have experienced a minimum loss of about 70% in mass (Jakosky et al. 2017). The low mass of the planet shall have played a role in such a massive loss, enhancing Jeans escape. Impact erosion by large bodies was another factor. Recent measurements by the MAVEN satellite suggest however that erosion by solar wind represented the single most important factor in the evolution of Mars, another object that lacks intrinsic magnetic field. MAVEN data detected spikes in the atmospheric escape during energetic plasma coronal mass ejections from the Sun. These events are believed to have occurred much more frequently in the early life of the Sun and may have therefore represented a major cause of atmospheric loss for Mars (Curry et al. 2017).

4.2.4 Photochemistry

The photo dissociation of atmospheric molecules by UV solar radiation represents a key factor in shaping the chemical cycles occurring in planetary atmospheres. In oxidative environments, such as the ones of rocky planets, the most important species is atomic oxygen, due to its high electro negativity. In reducing environments, such as the one found in giant planets, the dissociation of carbon and nitrogen bearing species (methane and ammonia respectively) and lack of strong oxidants allow the development of complex chemical patter and the production of a variety of heavier molecules. An extensive discussion on the subject is provided by Yung and de More (1999).

4.2.4.1 Venus and Mars

The photodissociation involves the main atmospheric component

$$\mathrm{CO_2} + h\nu \rightarrow \mathrm{CO} + \mathrm{O} \qquad \lambda < 2050\,\text{Å}\,. \tag{4.30}$$

In the thin Martian atmosphere, the reactions occur efficiently down to the surface. On the thicker atmosphere of Venus, the maximum production rate is expected at about 60 Km above the surface. The direct inverse reaction is extremely slow, since it requires a third molecule M to be involved, that is

$$CO + O + M \rightarrow CO_2 \tag{4.31}$$

Despite the very different rates of these reactions, the carbon dioxide mixing ratios remains much above the expected levels in the atmospheres of both planets, pointing toward the existence of other mechanisms to replenish the CO_2 content. In both cases, catalytic cycles involving minor components have been identified.

In the Martian environment, odd oxygen produced from the trace amounts of water vapour (again by photodissociation reactions) is involved in the reaction

$$CO + OH \rightarrow CO_2 + H\,. \tag{4.32}$$

Notably, the overall cycle balance is such that no net loss of water vapour occurs at the end of the cycle, where water acts therefore as a catalyst.

In the Venus environment, the recombination involves chlorine, a trace component confined in the lower troposphere because of the efficient condensation of chloride acid (the main Cl-bearing species) in the lower clouds:

$$\begin{aligned} &Cl + CO + M \rightarrow ClCO + M \\ &ClCO + O_2 + M \rightarrow ClC(O)O_2 + M \\ &ClC(O)O_2 + Cl \rightarrow CO_2 + ClO + Cl \end{aligned} \tag{4.33}$$

Venus Express measurements confirmed global scale patterns in the distribution of CO consistent with a creation at high latitude, equatorial locations and subsequent transport and destruction at lower altitudes, poleward positions.

4.2.4.2 Giant Planets

Methane is dissociated by photons with $\lambda < 1625$ Å. In the typical Jupiter conditions the photodissociation occurs at pressure levels with $p < 10^{-3}$ bar. Despite the low densities found there, the tendency of carbon to catenation results in a wide range of organic molecules being produced, most abundant being C_2H_6 (ethane) and C_2H_2 (acetylene). In Jupiter, despite air temperatures warm enough to forbid the methane condensation, a substantial depletion of methane occurs between the 10^{-7} and 10^{-8} bar levels, where concentration of photodissociation products is expected to peak. In the atmosphere of the icy giants, condensation of methane is expected to occur at levels much deeper that those affected by photodissociation. Therefore, the detection of organic molecules in the stratosphere of both Uranus and Neptune has been interpreted as a possible evidence of the occurrence of vertical motion transporting methane to its own dissociation levels.

Ammonia is also effectively photo dissociated by photons with $\lambda < 2300$ Å. In this case, an important product is represented by hydrazine

$$NH_3 + h\nu \rightarrow NH_2 + H \tag{4.34}$$
$$NH_2 + NH_2 \rightarrow NH_4$$

Since hydrazine is expected to experience prompt condensation in the conditions met in the upper atmospheres of giant planets, it is invoked as a key component of high altitude hazes observed in these environments (see also Sect. 4.4.3).

In the Jupiter atmosphere, the above reaction (Eq. 4.34) is expected to occur at much deeper levels (200–300 mbar) than the one affecting methane. This is due to the deeper penetration of the less energetic photons affecting ammonia as well as the substantial condensation experienced by ammonia in the upper troposphere. Nonetheless, occasional rises of ammonia caused by large storms at levels enriched (by downward diffusion) in acetylene are invoked as a scenario to create compounds containing nitrile (−CN), isonitrile (−NC), or diazide (−CNN) groups, possibly responsible of the colours observed for features such as the Great Red Spot (Carlson et al. 2016).

4.2.5 *Aerosols*

Without any exception, all atmospheres of the solar system in collisional regime ($R_n << 1$) contain—at least occasionally—aerosols. In the great majority of cases, these aerosols are formed by the condensation of atmospheric gaseous species, being often minor components.

The condensation of gases can occur when the partial pressure P_p of a gas exceeds the equilibrium pressure between the gaseous and the liquid/solid phase. The equilibrium pressure is usually approximated by the Clausius–Clapeyron relation

$$\ln P_{eq} = -\frac{L}{R_{gas} T} + C \, , \tag{4.35}$$

being R_{gas} the gas constant, T the air temperature, L the specific latent heat of vaporization/sublimation and C a constant. Both L and C are characteristic parameters of a gas. This formula immediately demonstrates how the condensation critically depends upon air temperature and partial pressure of the involved species, both factors being - at least at local scale - sensitive to an high number of different dynamical factors. The approximation in Eq. (4.35) is valid for temperatures well below the critical temperature of the considered species.

Albeit Eq. (4.35) describes the possibility for condensation to occur, more extensive treatment is needed to characterize the growth of the dimensions of aerosol particles. For liquid aerosols, *homogeneous nucleation* occurs when vapor directly condense to form droplets. On the basis of considerations upon Gibbs free energy (Salby 1996), it is possible to demonstrate that molecules tend to re-evaporate until the droplet reaches a critical radius r_c

$$r_c \propto \frac{2\,\sigma}{k_B\,T\,\ln\frac{P_p}{P_{eq}}}\,, \tag{4.36}$$

being σ the surface tension. Only when the critical radius has been exceeded, the droplets tend to increase by diffusion of vapour through the droplet boundary.

In *heterogeneous nucleation*, the vapour molecules condensates initially over other type of aerosols, reaching therefore much more easily the critical radius required to overcome the surface tension. The heterogeneous nucleation represents therefore the key mechanism to initiate aerosol condensation in actual atmospheres. Ionization induced by magnetospheric precipitation or cosmic rays can induce droplet charging and enhance substantially the condensation processes in the upper atmospheres. In matter of fact, above the critical radius, growth of droplets is substantially modified by the reciprocal collisions, that becomes the dominant factor in later stages of droplet growth.

Beside condensation products, other types of aerosols are typically found in planetary atmospheres. They include: volcanic ashes, the products of surface erosions (dust clouds on Mars and Earth), the hydrodynamic emissions along gases from surface (geysers on Enceladus and Triton) and meteoric dust.

The role of aerosols in the atmosphere physics can hardly be overestimated:

- their capability to reflect incoming solar radiation and to adsorb IR radiation may alter substantially the overall energetic balance of an atmosphere;
- the absorption/release of latent heat may alter the energetic balance at local level;
- the condensation and gravitational precipitation of aerosol modify substantially the vertical distribution of minor species;
- aerosol can act as effective catalytic sites for a number of atmospheric chemical reactions.

Moreover, emerging radiation field is modified substantially by aerosol scattering, with major implications on remote sensing methods.

4.3 Fundamentals of Atmospheric Dynamics

4.3.1 Main Drivers

Planetary atmospheres experience large scale motion of air masses. In the small bodies (rocky planets, Titan, Pluto, Triton), the main driver for motions is represented by the differential heating of surface and atmosphere induced by different latitudinal exposure to Solar irradiation. In the giant planets, the effects due to the dissipation of internal heat become more and more important toward the interior.

The wind patterns caused by pressure differences are however substantially modified by the planet rotation and more specifically by the need for air parcels to preserve

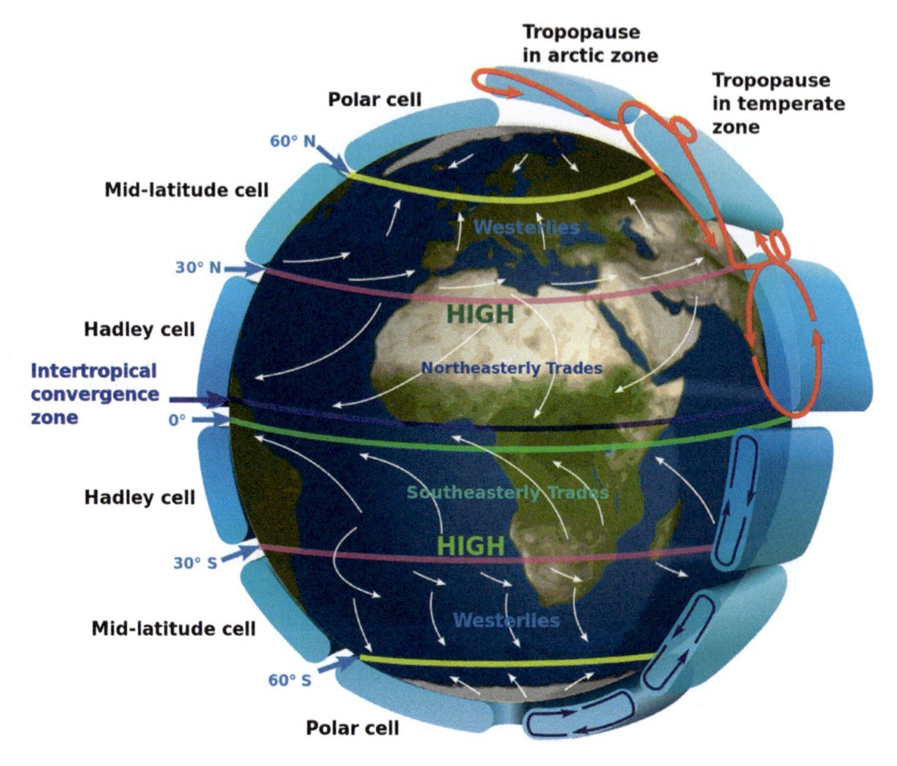

Fig. 4.3 General scheme of the General circulation on Earth troposphere. Courtesy NASA/Wikimedia Commons

their angular momentum. Other factors to take into account in assessing the global circulation of a body are:

- differential heating of surface induced by the contrast between land and oceans (on the Earth) or between high and low albedo areas (e.g. Mars);
- mass flow on condensible species (CO_2 on Mars, N_2 and CH_4 on Pluto and Triton);
- seasonal cycles, with particular attention to axial tilt and orbital eccentricity.

The discussion in this section largely follows the introduction provided by de Pater and Lissauer (2010). Further details are can be found in Salby (1996, Chap. 12).

4.3.2 Global Circulation Patterns

A common feature of atmospheric circulation in planetary atmospheres is the so-called Hadley circulation (Fig. 4.3). In the case of rocky planets, the heating of the surface at the subsolar latitudes leads to an expansion of air parcels, that become buoyant and rises in altitude, cooling during expansion and reaching levels where IR cooling become effective. During the rise, air tends to lose the condensible component to form aerosols: the release of latent heat further enhances the circulation. Once in altitude, the air parcels are subject to a net pressure gradient that leads them poleward. During this latitudinal motion, air parcels must preserve their angular momentum despite a lesser distance from planet rotation axis and therefore accelerates along parallels in the same direction of planet rotation. On the Earth this eventually leads to the formation of subtropical jets. At higher latitudes, air parcels (strongly depleted in condensible species) eventually sinks again toward the surface, heating by adiabatic compression. At low altitudes the air parcels are subject to a reverse pressure gradient (created by the ascending branch of the cell) and flows toward the sub solar regions. This surface flow must still preserve angular momentum and is therefore accelerated along parallels in the direction opposed to planet rotation.

The longitudinal extension of the cell is dependent upon temperature gradients at the surface as well as on the rotation speed. The very slow rotation of Venus and negligible axial tilt allow the Hadley circulation to develop in two symmetric, hemisphere-wide cells. A similar condition is found on Mars around equinoxes. On the Earth, the proper Hadley cell is extended approximatively up to latitudes of 30°. Another conceptually similar cell (polar cell) exist beyond 60°, similarly driven by surface temperature gradients. At intermediate temperate latitudes, the weak Ferret cell with an opposite circulation pattern can be found. This structure is essentially driven by the dragging of Hadley and polar cells at its boundaries, being its behavior (it transfers heat toward equator) thermodynamically adversed. Gaseous giants display similarly a large number of cells in both hemispheres, marked by strong jets at their boundaries. Structure of icy-giant circulation is known in a lesser degree, but is apparently characterized by a two larges cell in each hemisphere. It shall be stressed that meridional (i.e.: along meridians) and vertical motions associated to Hadley circulation are by far weaker than those associated to zonal (i.e.: along parallels) winds. On the Earth at the approximative level of 0.5 bar, vertical motions are usually well below the $1\,\mathrm{cm\,s^{-1}}$ value, to be compared to mean zonal winds up to $25\,\mathrm{m\,s^{-1}}$ at mid-latitudes.

Another type of global circulation, driven solely by solar irradiation, is the transfer of air from the warm sub-solar point to the anti-solar point on the night side, with a mechanism conceptually similar to the Hadley cell. This kind of circulation occurs in the upper atmosphere, where drag and irradiation from surface become negligible and UV solar radiation can induce a substantial deposition of heat. In the Venus case (Fig. 4.4), infrared observations has allowed to map directly the downwelling of UV dissociation products.

Another important feature related to global circulation is the dissipation of energy. In addition to the IR cooling mentioned earlier, an important role is played by the friction occurring in vicinity of surface in the planetary boundary layer. Here, turbulence may play an important role in transferring the energy from large scale eddies with scales comparable to the roughness of surface down to Kolmogorv scale length where molecular diffusion can dissipate it efficiently.

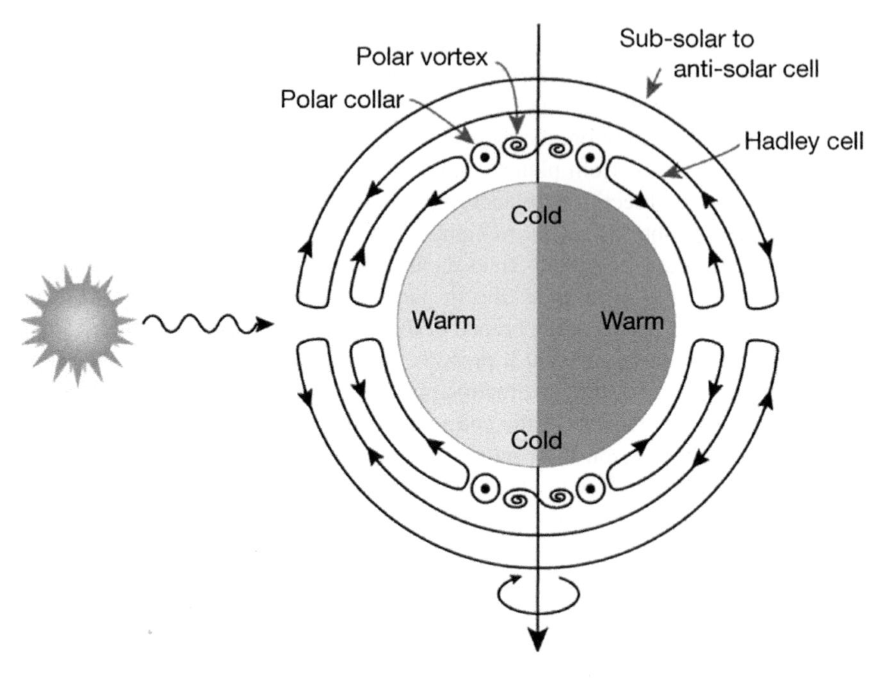

Fig. 4.4 General scheme of the Solar/Antisolar circulation observed in the Venus' upper atmosphere. Hadley circulation occurs simultaneously at altitude below $\approx$ 100 km. Reprinted by permission from Macmillan Publishers Ltd: Nature, Svedhem et al. (2007), Copyright 2007

4.3.3 Wind Equations

To describe numerically the winds, we start from the Navier-Stokes equation, Eq. (4.9), considering the rotating reference frame represented by planet surface. We denote the speed in this reference frame $\overrightarrow{u'}$. In this case

$$\rho \frac{D\overrightarrow{u'}}{Dt} = -2\,\rho\,\overrightarrow{\omega}_{\text{rot}} \times \overrightarrow{u'} + \rho\,\overrightarrow{g'} - \overrightarrow{\nabla}\,p - \overrightarrow{\nabla} \cdot \left(\mu \overrightarrow{\nabla} u'\right) \tag{4.37}$$

being $\overrightarrow{\omega}_{\text{rot}}$ the rotation angular speed of the planet. Naming as u, v and w the wind components along the meridians, parallels and the vertical direction, respectively, and neglecting viscosity we get:

$$\begin{aligned}
\frac{Du}{Dt} &= 2\,\omega_{\text{rot}}(v\,\sin\theta - w\,\cos\theta) - \frac{1}{\rho}\frac{\partial P}{\partial x} \\
\frac{Dv}{Dt} &= 2\,\omega_{\text{rot}}\,u\,\sin\theta - \frac{1}{\rho}\frac{\partial P}{\partial y} \\
\frac{Dw}{Dt} &= 2\,\omega_{\text{rot}}\,u\,\cos\theta - g - \frac{1}{\rho}\frac{\partial P}{\partial z} \approx 2\,\omega_{\text{rot}}\,u\,\cos\theta
\end{aligned} \tag{4.38}$$

where θ is the latitude and where, in the last formula, we have introduced the assumption of hydrostatic equilibrium, Eq. (4.12). In the so-called *shallow atmosphere approximation*, motions along the vertical directions are considered negligible on the basis of dimensional analysis (Salby 1996, Sect. 11.4.2) and, considering only the horizontal plane, we get

$$\frac{Dv}{Dt} = f_C \vec{v} \times \hat{z} - \frac{1}{\rho} \vec{\nabla} P , \tag{4.39}$$

where $\vec{v}$ is the speed vector on the horizontal plane, as seen in the rotating reference frame, $f_C \equiv 2\,\omega_{rot} \sin\theta$ is the Coriolis parameter and $\hat{z}$ the unit vector normal to the surface. Heuristically, the Coriolis term expresses the conservation of angular momentum of air parcels as they move between different latitudes.

A further important approximation can be made considering a balanced flow, where the Coriolis and pressure gradient terms compensate each other leading to a steady flow $(Dv/Dt = 0)$ in the *geostrophic approximation*. In these circumstances,

$$v = \frac{1}{\rho f_C} \left(\hat{n} \times \vec{\nabla} P \right) , \tag{4.40}$$

being $\hat{n}$ the unit vector normal to speed, implying that flow occurs along isobars. The jets forming at the descending branches of Hadley cells are the important examples of geostrophic balance.

If we introduce the geopotential Φ_g

$$\Phi_g = \int_0^Z g dz = - \int_{P_0}^P \frac{dP}{\rho} , \tag{4.41}$$

we can reformulate the previous discussion referring to isobaric surfaces and to restate Eq. (4.40) as

$$v = \frac{1}{f_C} \hat{n} \times \vec{\nabla}_P \Phi_g , \tag{4.42}$$

where index P indicates the gradient as computed over isobaric surfaces. Upon differentiation of Eq. (4.40) with respect to pressure and inserting Eq. (4.13) leads to

$$\frac{-\partial v}{\partial \ln P} = \frac{R_{gas}}{f_C} \hat{n} \times \vec{\nabla}_P T . \tag{4.43}$$

Namely, this *thermal wind balance* relates the latitudinal variation of temperatures with the vertical variations of zonal winds. Heuristically, in presence of a longitudinal temperature gradient, vertical spacing between isobaric surfaces tends to increase toward warmer areas, in a greater amount at greater altitudes. The net effect is an increased slope of isobaric surfaces with increasing altitude. At a fixed altitude, this implies an increased pressure gradient and the need of stronger winds to compensate the pressure difference. Thermal wind balance represents a common method to estimate winds (gradients) from maps of air temperature as a function of latitude and altitude away from equatorial regions when no direct wind measurements are available. Thermal wind is also the main driver of mid-latitude jets observed on the Earth (as explained above, sub-tropical jets are mostly driven by conservation of angular momentum at the descending branch of the Hadley cell).

Equation (4.39) can be separated in the normal and tangential components of the motion to achieve the following:

$$
\begin{aligned}
\frac{dv}{dt} &= -\frac{1}{\rho}\frac{d\rho}{ds} \\
\frac{v^2}{r} &= -f_C v - \frac{1}{\rho}\frac{dP}{dn},
\end{aligned}
\tag{4.44}
$$

being s the curvilinear coordinate along the motion and v the speed magnitude. This formulation allows one to consider balanced flow also in the cases where f_C is very small, such as in the vicinity of equator or in slow rotating planets. Namely, this *cyclostrophic balance* occurs when the pressure gradient along the direction normal to the flow is balanced by the centrifugal term. Important examples are the atmospheric super rotations observed in the atmosphere of Venus and Titan.

We can define the relative vorticity of a velocity field as

$$\omega_v = \vec{\nabla} \times \vec{u}\,, \tag{4.45}$$

The vorticity field expresses a measure of the tendency of the fluid to rotate around the point of interest. For an air parcel at rest with respect to the planet surface, it holds $|\omega_v| = 2\omega_{\text{rot}}$. Vorticity is particularly important in the description of eddies. Namely, common features of planetary atmospheres are represented by large vortices, where air tends to rotate around a local maxim/minimum of pressure. Cyclones are the vortices developing around a pressure minimum, while anticyclones are the vortices develop around a pressure maximum. The region where air is actively forced to rotate behaves approximatively like a rigid body, with tangential speeds proportional to distance from the center. Here vorticity has a non-zero value. On the outer parts, essentially dragged by the vortex core, the tangential speed tends to decrease as $1/r$, being r the distance from center. Here vorticity becomes zero. In cyclonic circulation, the *Rossby number* allows one to estimate if either *geostrophic* or *cyclostrophic* balance holds. It is defined as

$$R_0 = \frac{v}{L\, f_C}, \tag{4.46}$$

where v is the magnitude of the wind speed and L a characteristic scale of the phenomenon (such as vortex diameter). With $R_0 << 1$, the geostrophic approximations can be considered valid and pressure gradient is essentially balanced by the Coriolis acceleration. Small-scale systems (small L) or those developing at low latitudes are more easily dominated by centrifugal effects. With considerable simplifications, we can define the Rossby potential vorticity as

$$\omega_{pv} = \frac{\omega_v + f_C}{l}, \tag{4.47}$$

being l a quantity proportional to the thickness of the layer involved in the motion. Potential vorticity gives a measure of the angular momentum of fluid around the vertical axis. Once the turbulent dissipation occurring at planetary boundary layer is neglected, it can be demonstrated that potential vorticity of an air parcel must be preserved. A consequence of this requirement is that variations of latitudes (and hence of the Coriolis parameter) implies a variation of vorticity. The conservation of potential vorticity is also responsible for other phenomena such as the expansion of cyclones upon passing over topographic heights or their prevailing trajectories over Atlantic Ocean.

4.3.4 Atmospheric Waves

The overall global circulation patterns described in previous sections are rarely found in these basic forms in actual planetary environments. Conversely, these shall be considered as idealized dynamical schemes that emerges once highly-variable temporary features are removed by averaging over time. Atmospheres actually hosts a variety of wave features (periodic variations of the physical parameters of the atmosphere in time and space), that can often be treated mathematically as infinitesimal perturbations of the steady flow regimes. Here we briefly describe only a few types.

4.3.4.1 Thermal Tides

The thermal waves are created by the natural modulation imposed on solar input by the rotation of the planet (or by the possible super rotation of the atmosphere). Sun-radiation flux represents the effective forcing acting on the atmosphere. At a fixed location, its value versus time becomes abruptly null after the sunset (with a discontinuity in its first derivative). As typical of step-like functions, corresponding Fourier transform contains significant contributions from several harmonics of the fundamental period (in this case, the actual day length). More frequent periods are those associated to one day and half days. Most common tides appear as variations of air temperatures and altitudes of isobaric surfaces, being the variation phases approximatively fixed with respect to the sub-solar position and therefore seen as migrating by an observer on the surface. These includes, for example, the periodic variations of surface pressure clearly seen on Mars and air temperature minima/maxima locked at fixed local time positions on Venus.

4.3.4.2 Rossby Waves

Development of weather systems on the Earth mid-latitudes is often dominated by the so-called Rossby waves. Considering the Earth north hemisphere case, an air parcel, involved in the strong polar estward jet, can experience small longitudinal variations. Preservation of potential vorticity will modify its relative vorticity, representing an actual restoration factor: an initial displacement northward (decrease of Coriolis factor) will reduce relative vorticity, until it becomes negative and the air parcels effectively move back southward. Upon overshooting, the same potential vorticity conservation principle initially increases the potential vorticity, until it surpasses the initial value of Coriolis parameters and is deviated toward north. An observer moving westward would see the air to spin clockwise while moving north and clockwise while moving south. This mechanism creates the meander-like wind flow patterns often observed at mid-latitudes.

4.3.4.3 Gravity Waves

Gravity waves are vertical perturbations of the atmosphere where gravitation and buoyancy plays the role of restoration forces. Considering the case of an infinitesimal vertical displacement of the atmosphere, this can result in periodic oscillations at Brunt-Väisälä frequency

$$\nu_{\mathrm{BV}} = \sqrt{-\frac{g}{\rho_0}\frac{\partial \rho(z)}{\partial z}}, \tag{4.48}$$

being ρ_0 the air density at the initial level in the unperturbed atmosphere.

Equation (4.48) leads to a real value of the Brunt-Väisälä frequency—i.e. to periodic oscillations—only when density decreases with altitude. In the opposite case, perturbations of the system are amplified exponentially—i.e. the atmosphere is gravitationally unstable. Gravity wave are typically induced by a flow upon a topographic discontinuity, and as such has been observed on the Earth, Mars and—unexpectedly—Venus. They can however develop also in absence of topographic discontinuity (notably in giant planets) in a large range of instabilities in the stratification of the atmospheres.

4.3.5 Diffusion

In Sect. 4.3.2 we briefly described the motion of air parcels. The turbulence, usually found at the lowest atmospheric levels, ensures that atmosphere remains well mixed, i.e. with an uniform composition over altitude. At low Reynolds numbers R_e, Eq. (4.11), however, *molecular diffusion*, driven by the random motion of individual molecules, can create substantial deviations from the well-mixed conditions.

Namely, the motion of the molecules of a given chemical species is such to remove any gradient in its density as well in the temperature field. Moreover, given the dependence of Eq. (4.14) on molecular mass, lighter species tends to exhibits higher values of scale height H_i and to be more uniformly distributed in altitude with respect to heavier ones. At the same time, the eddy diffusion by turbulence tends to remove any compositional gradient. Quantitatively, the flux Φ_i of the molecules of the generic chemical species i across a horizontal surface is given by

$$\Phi i = -N_i D_i \left(\frac{1}{N_i} \frac{\partial N_i}{\partial z} + \frac{1}{H_i} + \frac{\alpha_i}{T(z)} \frac{\partial T(z)}{\partial z} \right) - N\, K \frac{\partial (N_i/N)}{\partial z}, \tag{4.49}$$

being N the total molecular number density, N_i the molecular number density of the species i. D_i is the molecular diffusion coefficient for i (in turn, inversely proportional to N) while K is the eddy diffusion coefficient (roughly proportional to R_e). This latter term is such to remove any gradient in the mixing ratio of the species.

4.4 Atmospheres of Individual Solar System Bodies

Our review of atmospheres of the Solar System considers the bodies that own significant parts of their atmosphere in collisional regime

4.4.1 Rocky Planets

Rocky planets are characterized by solid surfaces composed mostly of silicates. Interaction with the overlying atmospheres takes place mostly in the form of release of volcanic gases (Earth and likely Venus), molecular diffusion trough the soil and deposition/sublimation of polar caps (Earth and Mars). Moreover, surfaces represent the lower boundary of atmospheric circulation and the site of important chemical reactions involving atmospheric components. Table 4.3 summarizes the key properties of the atmosphere of rocky planets in the Solar system.

4.4.1.1 Venus

The most striking feature of the Venusian atmosphere is represented by the thick coverage of clouds that permanently precludes the visual observation of the surface. A state-of-the-art review on our current knowledge on Venus aerosols is provided by Titov et al. (2016).

Moving from the space toward the Venus surface, we firstly meet a population of sub-micron hazes (the so-called mode 1 particles), detected from an indicative altitude of about 100 km. A second, larger component ($r_m = 1.2\,\mu$m) is evident

Table 4.3 Properties of the Solar-system rocky planets that are more relevant for the discussion on their atmospheres. Taken from the NASA Planetary Fact Sheets (https://nssdc.gsfc.nasa.gov/planetary/factsheet/)

Parameter	Unit	Venus	Earth	Mars
Mass of the planet	$M_{\oplus}$	0.815	1	0.107
Surface gravity	$g_{\oplus}$	0.90	1	0.37
Magnetic field	Gauss cm^3	Weak, induced	Intrinsic, dipolar 8×10^{25}	Crustal, localized
Orbital semi-major axis	au	0.72	1	1.52
Sideral period	days	243	1	1.025
Surface pressure	bar	92	1	6×10^{-3}
Axial Tilt	degree	177.36	23.45	25.19
Main components				
CO_2	Volume fraction	0.965	400 ppm	0.956
H_2O	·	–	Variable (up to 5%)	–
N_2	·	0.035	0.78	0.019
O_2	·	Traces, upper atm.	0.21	0.014
SO_2	·	120 ppm (variable)	–	–
He	·	12 ppm	5.2 ppm	–
Ne	·	7 ppm	18.0 ppm	2.5 ppm
Ar	·	70 ppm	9340 ppm	16000 ppm

below 70 km. This component (mode 2) is the main constituent (in terms of mass) of the upper clouds of Venus, which represent what is typically observed from space in visible and infrared. A local minimum in cloud opacity at roughly 57 km marks the transition at the middle/lower cloud deck, where larger particles ($r_m = 3{-}5\,\mu$m, mode 3) are found. The altitude of 48 km sees a sharp decrease in aerosol opacity, and below only much optically thinner diffuse haze and possibly discrete clouds of uncertain nature can be found. In situ analysis, as well as remote IR and polarimetric measurements, have allowed us to identify a liquid mixture of sulfuric acid and water as the main constituent of haze and upper clouds: consistently, the clearing observed at the altitude of 48 km is met where temperature in the Venus environment allows the sulphuric acid to evaporate. Other constituents must be present as well: UV observations show high contrasting details, demonstrating the existence of a still unidentified UV absorber, strongly variable in space and time; the VEGA balloon instruments clearly demonstrated the importance of chlorine, phosphorus and iron in the deeper clouds, to form still unidentified components.

The long observation campaign of Venus Express from 2006 to 2014 represents a milestone in the exploration of the Venusian atmosphere. The large suite of instruments, operating from thermal IR (5 μm) to UV, enables a series of studies, mostly focused on upper clouds and hazes. Among the several results, we can cite the following. The studies of longitudinal trends in upper-cloud heights and scale heights demonstrated that both parameters decrease poleward from about 50° in longitude. The Sun-occultation measurements demonstrated that hazes occasionally present detached layers and they are characterized by a bi-modal size distribution, making therefore weaker the distinction between hazes and upper clouds. Further studies on phase functions derived from VMC nadir observations show that sub-micron particles are preferentially found on the morning hemisphere and highlighted an increase of refractive index in the region between 40° S and 60° S, as well as a slight increase of the size for mode 2 population toward the poles.

The sulphuric acid composing the clouds is created by the reactions

$$\begin{aligned} SO_2 + O &\rightarrow SO_3 \\ SO_3 + H_2O &\rightarrow H_2SO_4 \end{aligned} \tag{4.50}$$

being the atomic oxygen in the first reactions provided mostly by photodissociation of water vapour. The sulphur dioxide is assumed to be a product of volcanic activity. Albeit no firm evidence of on-going volcanic eruptions has yet been find on Venus surface, long term variations (in the order of several years) of SO_2 content in the upper troposphere confirm that ultimate sources of this compound shall be extremely irregular.

The number densities of different aerosol size populations vary rather continuously in altitude, but we can state that Venus clouds observed in the visible are essentially those associated to mode 2 particles and that we reach an optical depth equal to 1 (used as effective measure of cloud altitude) at roughly 73 km above the surface. Visible observations are rather featureless and only the UV absorbers allow one to track effectively the atmospheric motions at the cloud top (Fig. 4.5).

Venus atmospheres displays an atmospheric super rotation with winds at about $100\,\mathrm{m\,s^{-1}}$ up in the latitude range 20°–50° on both poles, following roughly the profile expected from cyclostrophic winds (Fig. 4.6). The equatorial region is not longitudinally uniform: the locations at local times between approximately 12 and 16 are rather irregular, and are interpreted as being subject to intense local convection. The morning side appears darker, and darker regions extends also to afternoon local times at longitudes between 30° and 50°. Darker regions are usually interpreted as evidence of longer residing times of air parcels at cloud top. Immediately poleward of the 50 parallels, the zonal speed begins to decrease, while the much weaker meridional winds show a local maximum (max $10\,\mathrm{m\,s^{-1}}$). At roughly 50° latitude, the cloud top altitude begins to decrease steady toward the polar value of about 65 km. At the same location, a bright polar collar is usually observed, embracing a darker area centred at about 70°.

Night time observations at the CO_2 transparency windows in the near IR allow one to map the relative depth of the cloud deck and to monitor, in more transparent latitudes, the winds at the approximate level of 50 km (Fig. 4.7). Latitudinal profiles of zonal winds still present a flat region between −50° and 50°, but magnitudes are

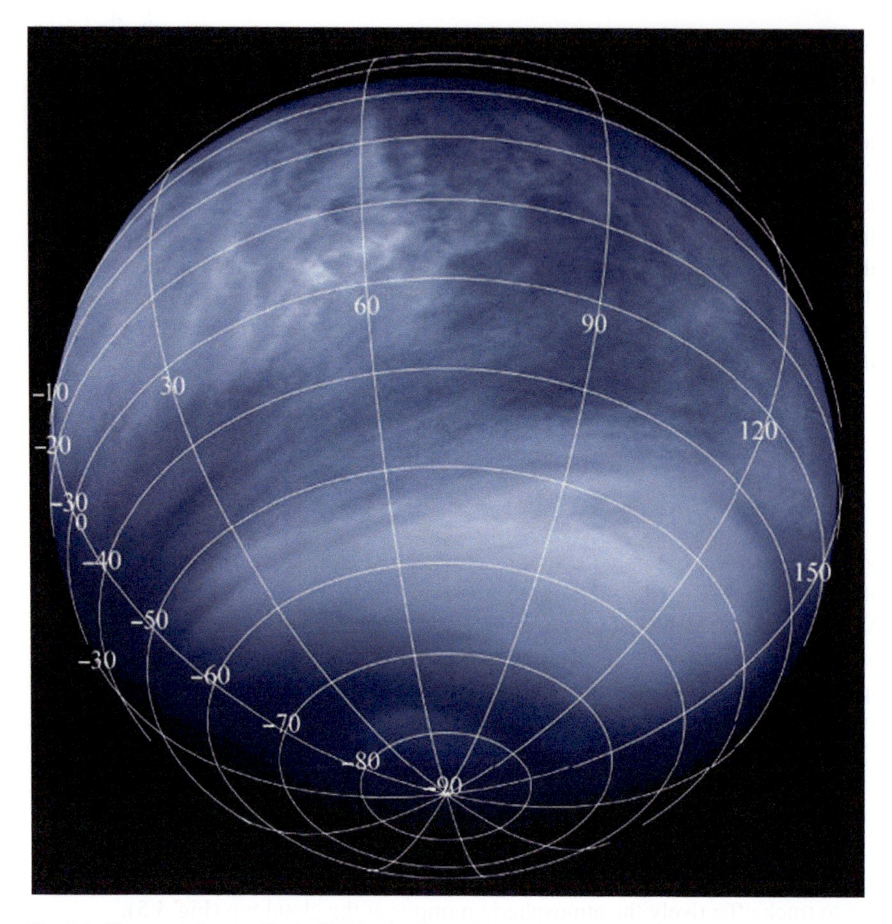

Fig. 4.5 Venus as observed in the UV by the Venus Monitoring Camera (VMC) on board of Venus Express. Reprinted by permission from Macmillan Publishers Ltd: Nature, Titov et al. (2008), Copyright 2008

reduced to about 70 m s^{-1}. Data from ground probes (landed at a maximum longitude of 60°) suggest indeed that wind speeds decrease steadily toward the surface, where no detectable winds were measured. This apparent lack of surface winds however imposes considerable difficulties both in justifying the closure of Hadley circulation as well as in explaining the gravity waves associated to the topographic relief of Aphrodite Terra recently detected by the Akatsuki spacecraft (Fukuhara et al. 2017).

Direct measurements by entry probes allowed us to infer a surface temperature of 740 K and, for the air temperature gradient below the 50 km level, a value very close to the adiabatic profile. Here convection is indeed the dominant form of energy transport from deep atmosphere to outer layers, since the very high IR opacity, caused by the thick CO_2 atmosphere, keeps the radiative cooling rate very close to zero up to an altitude of 60 km. A large body of data has been collected about the thermal

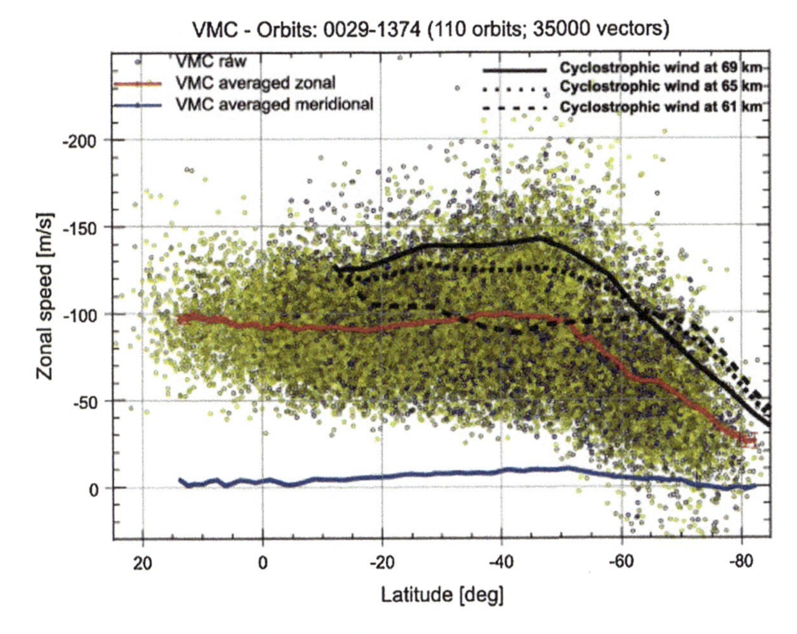

Fig. 4.6 Venus zonal and meridional wind speed as determined from VMC UV measurements. Reprinted from Piccialli et al. (2012), Copyright 2012, with permission from Elsevier

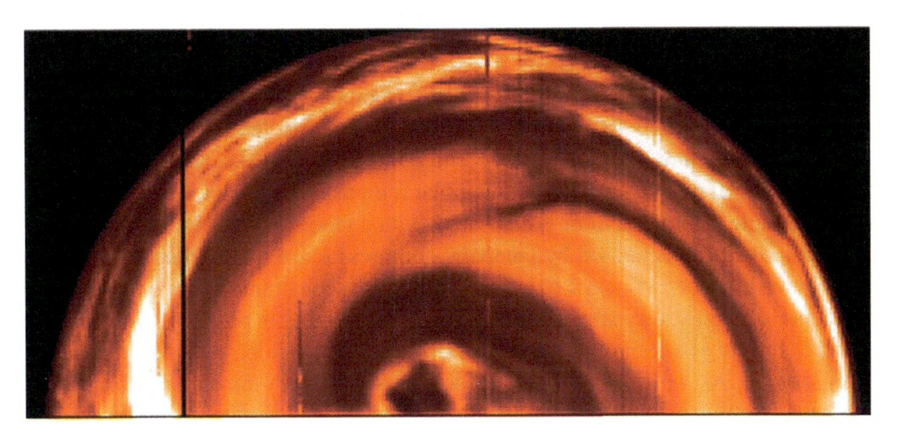

Fig. 4.7 Southern hemisphere of Venus as observed over the night side by the Visual and Infrared Imaging Spectrometer (VIRTIS) on board of Venus Express. The image was acquired in the CO_2 transparency window of 1.7 μm. Courtesy ESA/INAF/Obs. De Paris

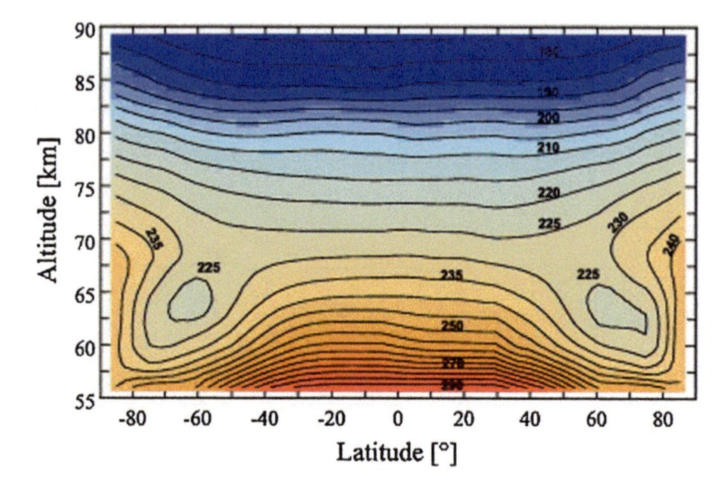

Fig. 4.8 Mean air temperature in the Venus night-time mesosphere, as a function of altitude and longitude. Reprinted from Haus et al. (2014), Copyright 2014, with permission from Elsevier

structure of Venus atmosphere above the 55 km level by the Venus Express instruments (Limaye et al. 2017). The temperature field exhibits a remarkable degree of symmetry between the two hemispheres, as expected from the very small axial tilt. At level around 55 km (Fig. 4.8), air temperature tends to decrease monotonically from the equator to the poles (from 290 to 240 K), as expected by intense absorption of UV radiation by clouds at the sub-solar points. Already at the 65 km level, dynamical effects begins to dominate, with two symmetric "cold collar" showing a minimum temperature of 220 K at 65° S and 65° N. Above 70 km and at least up to 90 km, equatorial regions are colder than polar regions, with monotonically increase at fixed altitude. This behaviour is consistent with a global scale Hadley circulation, where polar heating is caused by adiabatic compression of descending air. The air temperatures between 90 and 110 km are still poorly constrained by available data, that seems to suggest very strong variations. This variability is possibly related to the overall change in global circulation pattern (from Hadley to solar/antisolar) occurring at this altitude. Above 120 km, the day side atmosphere is dominated by absorption of UV radiation by molecules, with locations of peaks values of temperature strictly following the subsolar point. The solar/antisolar temperature gradient around 120 km was found to be around 35 K.

Several transient phenomena have been observed in the Venus atmosphere, at different time scales. Beyond the decade-scale variations of SO_2 content, variations over several years have also been reported for the cloud altitudes over the poles, and in the same region, for global brightness of polar regions in UV and visible. In the IR images, both poles, well within the cold collars, host bight (i.e. warm) dipole-shaped structures (Fig. 4.9). These structures show a complex variability on the scale of a few days and are considered, along with the outer cold collars, as polar anticyclonic systems embedded in the overall Hadley circulation. Notably, the entire region poleward of 60° S shows variability in the spatial distributions of air temperatures on the scale of few days, with patterns compatible with cyclonic arms, at altitudes up to 70 km.

4.4.1.2 Mars

Our knowledge of Martian atmosphere benefits of the high number of missions that have targeted this planet from the dawn of space era. The Martian environment is unique is several aspects, most important being the high axial tilt (25°) and orbital eccentricity (0.09)—leading to strong and hemisphere-asymmetrical seasonal cycles—as well as the substantial fraction of the atmosphere involved in the seasonal sublimation/condensation processes.

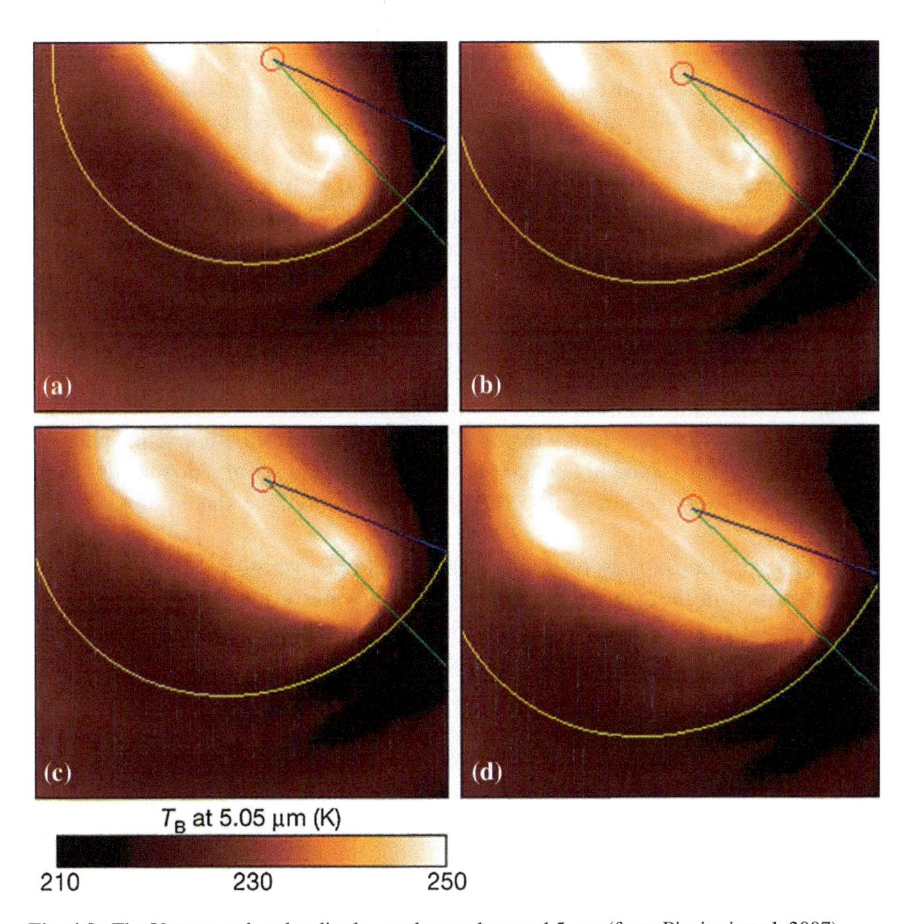

Fig. 4.9 The Venus south polar dipole, as observed around 5 μm (from Piccioni et al. 2007)

Considering the solar longitude L_S as a measure of season (being $L_S = 0$ the northern spring equinox), the perihelion occurs at $L_S = 250.87$, during the southern summer. This leads to a net difference of 63% in surface solar flux during the respective summer seasons between the two hemispheres. During polar night, on both hemispheres, the surface temperature falls below the condensation temperature of carbon dioxide and the gas, the principal component of atmosphere, begins to condensate over the surface. During southern winter, up to 25% of the atmospheric mass can be removed from atmosphere and up to 10% during northern winter, leading to comparable variations in surface pressure over the entire planet. These amounts correspond to the thicknesses of few meters for the seasonal slabs forming over each pole during winter. However, both polar caps are mostly formed by water ice, that forms permanent deposits with thicknesses in the order of a few kilometers.

The Martian surface has been prone to substantial erosive action along its geological history and shows a silicate-rich, fine-grained texture over most areas. Despite the small surface air density, wind shear is enough, in several circumstances, to lift micron-sized dust from the surface and to create dust storms. These phenomena may develop at local scale as dust devils of few hundreds meters of altitudes. Dust devils are more frequent in the warmest hours of the afternoon, when surface warming by the Sun may induce instabilities in the planetary boundary layer. More important from the global climate perspective are however the global dust storms, that engulf large fraction of the planet (Fig. 4.10 and top panel of Fig. 4.11). While at peak of dust storm activity the 10 μm opacity exceeds often the value of 0.5, even in the most clear periods it does not usually fall below 0.1, leaving therefore permanently a dusty background in the Martian sky. Global storms develop preferentially in the late summer of the southern hemisphere, and display a remarkable year-to-year variability. Minor storms are also observed during the northern summer, but with lesser total load (up to 0.3). Rather interesting are also the sporadic increases of dust observed at the rims of both polar caps during the late spring of the corresponding hemispheres. These events are associated to the mass flows related to the sublimation of the polar caps and has allowed us to track directly the development of weather systems in sub polar regions.

The Martian atmospheres host also water ice clouds, whose phenomenology is better described in the general context of Martian water cycle. The current mean surface pressure of Martian atmosphere (6 mbar) lies very close to the pressure of water triple point and consequently water can not exist permanently on the surface in liquid form. Nonetheless, several geological evidences (including dry river beds, sedimentary rock forms, minerals created only in wet environments) demonstrates that substantial bodies of liquid water must have existed—at least sporadically—over the planet surface in a remote past. At the current date, water is found in three main reservoirs: atmosphere, polar caps and soil. Water vapour displays a clear seasonal cycle (Fig. 4.11, lower panel), with concentration peaks observed at the rim of retreating polar caps of each hemisphere during the corresponding late spring ($L_S = 110, 85°$ N and $L_S = 290, 80°$ S). The evolution of water vapor content against L_S and latitude is generally consistent with a sublimation from polar caps becoming warmer with increasing isolation and subsequent transport toward equator along

Fig. 4.10 Onset of a Martian global dust storm as observed by the Hubble Space Telescope. Courtesy NASA/J. Bell (Cornell), M. Wolff (SSI). STScI release PRC01-31

the near-surface Hadley circulation. However, it was already noticed from Viking measurements in late '70 how—during the northern spring—water vapour starts to increase almost simultaneously at $L_S = 45$ on a large range of latitudes. This imply a release not just from the polar cap, but also from a soil-related water reservoir at the mid-latitudes. Full evidence of the importance of these reservoirs for water was gained by the usage of gamma ray and neutron spectrometers on board of Mars Odyssey. These data show that water-equivalent hydrogen (likely in the form of ice) soil mass percentages up to 10% can be found even at equatorial latitudes. Seasonal cycle of water vapor shows an obvious hemispheric asymmetry, with substantial releases from northern cap, subsequent transport toward equator and—at least in some years—further transport toward southern hemisphere (visible as slight increases at $L_S = 210$, 30° S). Release from southern cap (that remains much colder that its northern counterpart during respective winters) is much more limited, and no obvious equator-ward water transport is evident.

Water ice clouds (Fig. 4.11, middle panel) are often observed over both poles during the corresponding cold seasons, with a typical particle size of 1 μm. Another important feature is represented by the increase of water clouds over equatorial regions occurring after the aphelion, in the late northern spring (aphelion cloud belt).

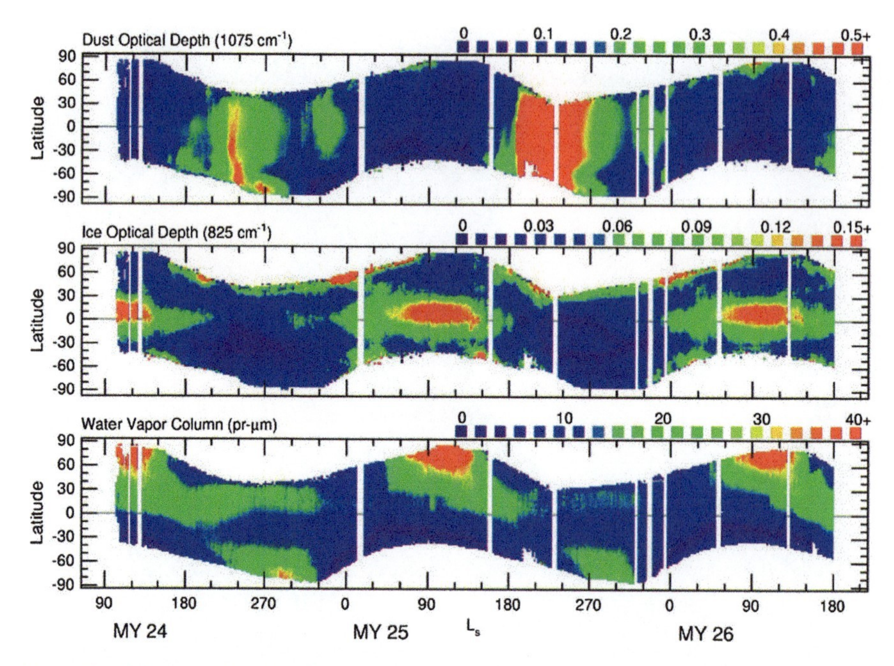

Fig. 4.11 Multi-annual trends of longitude-averaged values of dust optical depth, water ice optical depth, and water vapour content as a function of season (as quantified by the L_S parameter) and latitude. The values were retrieved from the data of the Thermal Emission Spectrometer on board of NASA Mars Global Surveyor satellite and refers to 14 local time. Reprinted from Smith (2004), Copyright 2004, with permission from Elsevier

These clouds tend to persist for rather long time in the atmosphere, as demonstrated by their relatively larger particle sizes (modal radius of 2−3 μm). It is yet not clear if the water mass involved in the aphelion cloud belt could compensate the hemispheric asymmetry in the water vapor cycle or, contrary, we are observing a long-term net transfer of water from the north to the south hemisphere. Water ice clouds have also been reported as orographic clouds in the vicinity of great volcanic domes and over the morning limb. Moreover, data from surface landers demonstrated the existence of a daily water cycle, characterized by formation of brines and low altitude clouds during the night and subsequent sublimation in the early morning hours.

A final important type of aerosols in the Martian atmosphere is represented by carbon dioxide clouds. These clouds occur at mesospheric altitudes (usually above the 60 km levels) and form preferentially at equatorial latitudes (20° S–20° N) between northern late winter ($L_S = 330$) and mid-summer ($L_S = 150$) with a pause around summer solstice. Sporadic detections were reported up to 50° latitude on both hemispheres (Määttänen et al. 2010). The formation of these clouds has been related to the creation of supersaturation pockets by interference of thermal tides and vertically-propagating gravity waves.

Ozone on Mars is produced by the recombination of oxygen atoms created by photolysis of carbon dioxide in the upper atmosphere over the dayside of the planet (Montmessin and Lefèvre 2013). Descending branch of the Hadley circulation transports the oxygen-rich air to the polar regions, where upon higher densities, recombination can take place. Ozone is effectively destroyed by hydrogen radicals generated by the water photodissociation. Ozone shows indeed a peak concentration over the winter south pole, with significant concentration at ground levels (0.5 ppmv) and in a distinct layer at an altitude of 50 km (3 ppmv). During southern summer, another mid-altitude layer is observed over the equator (up to 2 ppmv), while the northern polar winter does not exhibit any high-altitude layer but only a surface enhancement of smaller extent than its southern counterpart (0.3 ppv). The observed concentrations and hemispheric differences were quantitatively modeled considering the more vigorous Hadley circulation during southern winter, capable to transport higher amounts of water in the upper atmosphere, where it experiences photodissociation and eventually provides an effective source of hydrogen radicals for the northern winter polar region.

Methane on Mars has been subject of heated debates over the last decade, because the possible implications on biological sources. Initial ground-based and remote sensed detection claims already indicated a very low concentration (10 ppbv), extremely variable in time and space. These measurements were however affected by considerable uncertainties and were regarded with suspicion until the Curiosity rover, after being initially unable to detect the compound above the uncertainty level, recorded a sudden increase and provided eventually a firm evidence of methane occurrence on the Martian surface (Webster et al. 2015). Four estimates in a two months period provided values around 8 ppbv (with an uncertainty of 2 ppbv), but later measurements fall again below the uncertainty level. Methane is effectively dissociated by UV radiation in the Martian environment, but the photodissociation lifetime is not fast enough to justify the rapid oscillations seen at global scale. Possible additional photochemical removal mechanisms include reactions with hydrogen peroxide, a product of water photochemistry. Sources must also act on short time scales: a hypothesis worth to mention is based on the episodic release of methane from geological deposits (clathrates dispersed in the ice-rich soil) formed by the interaction of liquid water with silicate rocks in the remote past of the planet ('serpentinization').

Temperature structure of Martian atmosphere has been studied in detail in its latitudinal, local time and seasonal variability by a number of remote-sensing instruments. A complete picture is provided by McCleese et al. (2010) (Fig. 4.12). At the northern spring equinox ($L_S = 0$), the atmosphere exhibits two hemisphere-wide symmetric Hadley cells, with moderate warming over both poles caused by the compression of the descending branch of the cell. As season proceeds toward northern summer solstice ($L_S = 90$), the southern cell becomes more and more vigorous, while its northern counterpart fades away, moving toward the condition of a single, planet-wide cell where polar temperature inversion (regions of positive vertical lapse rate) over north pole eventually disappear while vertical and longitudinal gradient over the south pole becomes maximum. Temperature field regains its approximate hemispherical symmetry around the northern autumn equinox ($L_S = 180$). Southern summer follow qualitatively a similar evolution, but due to orbital eccentricity, sub-solar warming at solstice is greater and air temperatures correspondingly higher. The maximum differences between day and night temperatures are observed below 20 km and above the 50 km level. The prompt response of the atmosphere in this latter region is justified by the effective absorption of Solar photons in the most opaque CO_2 regions in near-IR. The behavior in the lowest atmosphere demonstrates the strong coupling with surface temperatures. This coupling has been investigated in details by the Mini-TES instruments on board of Spirit and Opportunity rovers (Smith et al. 2004). In particular, their dataset demonstrated the quick rise of surface temperatures just after the sunrise and the development of convection in the lowest kilometer already around 10 LT. During nighttime, surface temperatures falls quickly after sunset, leading to the development of strong temperature inversions in the first kilometer above surface. In the lowest 40 km above the surface, scenario can be considerably complicated by the occurrence of dust storms. Silicate dust is an effective IR absorber

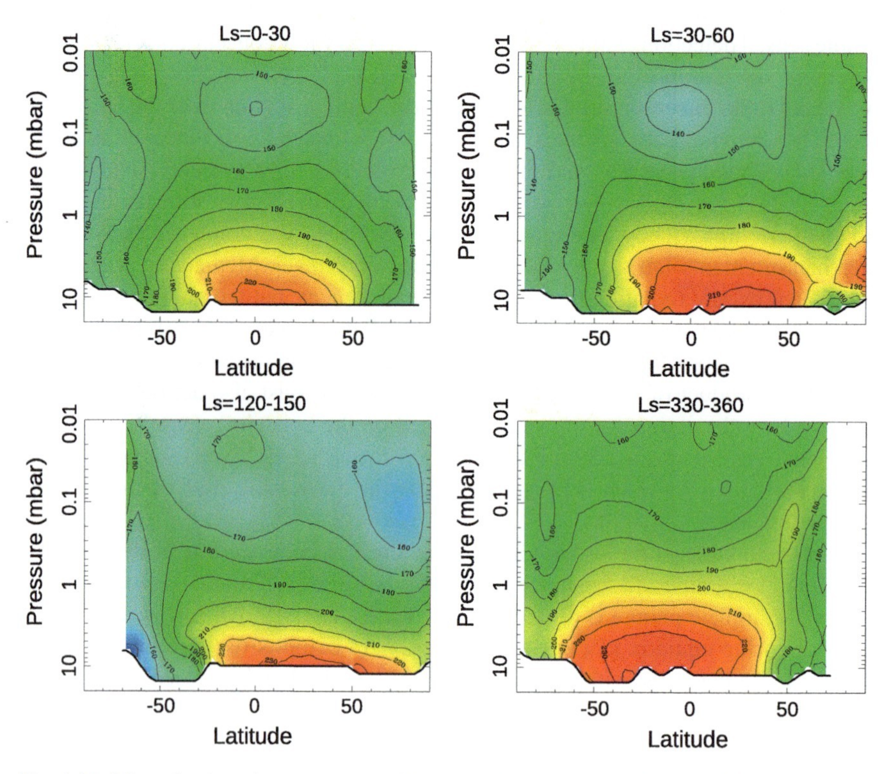

Fig. 4.12 Mean daytime air temperatures (Kelvin scale) in the atmosphere of Mars in different seasons, as a function of pressure and latitude. The values were retrieved from the data of the Planetary Fourier Spectrometer onboard the Mars Express satellite. Courtesy M. D'amore / DLR. See also McCleese et al. (2010) for a more complete coverage

and therefore its occurrence tends to rise the air temperature at levels up to 40 km, by absorption of the radiation thermally emitted by the surface and by absorption of solar visible and near IR radiation. On the other side, for the same reasons, it reduces the solar flux on the surface, with the net effect to reduce the thermal gradient in the lowest parts of the atmosphere.

4.4.2 *Giant Planets*

Contrarily to rocky planets, giant planets present a bulk composition with a substantial amount of light atoms, notably hydrogen and helium. This results into massive atmospheres that extend in altitude for a several thousands of kilometers, with gradual increase of the pressure and temperature toward the center. The lack of a shallow lower boundary interface for the atmosphere complicates considerably the study of these bodies, both from a theoretical as well as from an experimental perspective.

From a theoretical perspective, one shall note that circulation and aeronomy occur in principle over an extremely large range of different conditions of pressure and temperature, strictly mutually dependent.

From an experimental perspective, remote sensed data are capable to provide, at the very best, information about the first 100 km below the reference 1 bar pressure in the microwave domain. Constrains on the deeper structure has to be inferred rather indirectly from considerations on the basis of mass and radius or from detailed structures of gravity and magnetic fields (to be measured by spacecraft in the close vicinity of the body) with methodologies derived from geophysics of rocky planets rather than from atmospheric science. Table 4.4 summarizes the key properties of the atmosphere of giant planets in the Solar System. A comprehensive description of these atmospheres is provided by Irwin (2009).

4.4.2.1 Jupiter

Jupiter is the most massive giant planet and, in several meanings, the archetype of this class of bodies (Bagenal et al. 2004). For a visual observer, the most striking feature is represented by its complex cloud coverage. Given the high content of hydrogen, different elements are present in the Jupiter atmosphere in their reduced forms. Among these, water, ammonia and hydrogen sulphide are expected by thermochemical models to produce several decks of different composition between −70 and +20 km around the 1-bar level, with particle sizes in the order of a micrometer (Fig. 4.13). Moving upward from the interior, we meet first conditions where a solution of ammonia and water solution can exist in liquid form. The vertical extent of this cloud is strongly sensitive to the actual content of ammonia and water: indeed, upon further cooling of the atmosphere water forms water ice clouds and ammonia remains in gaseous form. Higher in the atmosphere ammonia combines with hydrogen sulphide to create ammonium hydrosulphide. Eventually, in the upper parts of

Table 4.4 Properties of giant planets more relevant for the discussion on their atmospheres. Updated from NASA Planetary Fact Sheets (https://nssdc.gsfc.nasa.gov/planetary/factsheet/)

Parameter	Unit	Jupiter	Saturn	Uranus	Neptune
Mass of the planet	$M_{\oplus}$	317.8	92.5	14.6	17.2
Magnetic field[a]	Gauss cm^3	20000	600	50[b]	25[b]
Orbital semi-major axis	au	5.2	9.5	19.2	30.0
Sideral period	days	0.41	0.43	0.72	0.67
Axial Tilt	degree	3.12	26.73	82.23	28.33
Main components in the observable troposphere					
H_2O	Volume fraction	0.86	0.87	0.82	0.80
He	·	0.13	0.11	0.15	0.19
CH_4	·	1.8×10^{-3}	4×10^{-3}	0.02	0.015
NH_3	·	7×10^{-4}	1×10^{-4}	?	?
H_2O	·	5×10^{-4}	?	?	?

Notes
[a]Magnetic field at magnetic equator (Earth = 1).
[b]Significant higher-order components are present in the global magnetic field

the troposphere, remaining ammonia is expected to form ammonia ice clouds. It shall be noted however that clear spectral detections of pure materials in Jupiter clouds are still sparse, suggesting therefore that substantial amounts of contaminants shall be present. The zones (bright) and belts (dark) pattern typical of Jupiter (Fig. 4.14) has a direct relation with cloud structure. The bright zones are interpreted as regions of vigorous air upwelling. Here, minor components from the deeper levels are effectively transported upwards by strong vertical motions. Upon cooling, these volatiles condense to form thick clouds. Since these materials are freshly provided by upwelling, they include only limited amounts of photodissociation products and clouds retain therefore in large extent their bright colour. Descending branches of this circulation are represented by the belts, where the air, depleted in volatile components but enriched in products by photodissociation (due to longer exposure time to sunlight) are warmed by adiabatic compression. In these conditions, cloud formation occurs in a lesser extent and is enriched in darker material. This scenario is qualitatively confirmed by the IR observations: at 5 μm the optical depth equal to one is reached at about 60 km below the 1-bar level, due to the H_2 absorption induced by collision between with others H_2 molecules or He atoms. Thermal emission from this 'deep' atmosphere is blocked by the overlying clouds over the zones, while belts appears much brighter, indicating eventually substantially thinner clouds. Available evidence suggests that in most circumstances IR and visible observations refer to clouds putatively composed by ammonium hydrosulfide and ammonia, with only sparse and rather indirect indications of the water clouds in the zones. This is indeed not surprising, since these decks are usually expected to be masked by the thick upper layers.

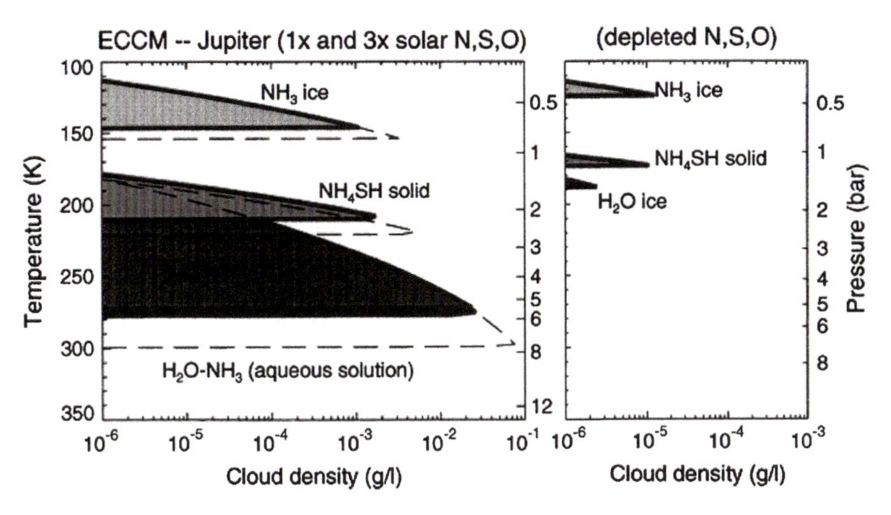

Fig. 4.13 Theoretical profiles of aerosol concentrations in the atmospheres of Jupiter. Left-hand panel presents the different expectations for solar composition (solid curved) and for an enrichment of a factor three with respect to hydrogen (dashed lines). Right-hand panel presents the expectations for an atmosphere depleted in volatiles (e.g. hot spots). Reprinted from Atreya et al. (1999), Copyright 1999, with permission from Elsevier

Fig. 4.14 True-color simulated view of Jupiter as reconstructed from Cassini Imaging Science Subsystem. Traditionally, bright latitudinal regions are indicated as zones, darker latitudinal regions (within 70° on both hemispheres) as belts. NASA Photojournal PIA02873. Courtesy NASA/JPL/University of Arizona

Above the upper cloud deck, a diffuse, sub-micron haze have been detected up to 200 mbar. The haze layer appears higher over the equator and has opacities in the visible well below 0.1. These hazes are likely formed by complex organic compounds, in turn produced from the photodissociation of methane and ammonia (as discussed in Sect. 4.2.4.2). A distinct, stratospheric haze population has been detected over the polar regions, where dissociation of atmospheric components is dramatically enhanced by the impinging of charged energetic particles in auroral regions. It is assumed that products of photodissociation are eventually removed from upper atmosphere by downward diffusions to higher pressure levels, where more efficient convective overturning performs an effective transport to the deep troposphere. Here, higher temperatures and pressures are effective in dissociating the more complex molecules and create back methane, ammonia and water upon reaction with hydrogen.

Jupiter clouds systems allow one to track one of the most complex dynamic observed in the Solar System. In absence of a solid surface, winds are usually refereed to a coordinate system (System III) defined on the basis of regular variations in radio emissions, in turn caused by rotation of Jupiter magnetic field, assumed to be representative of the rotation in the deepest parts of the planet. Similar coordinates systems have been defined for all giant planets.

On Jupiter, zonal winds show a clear pattern, marked by jets at the boundary between the zones and the belts (Fig. 4.15). Prograde jets are much stronger (up to $140\,\mathrm{m\,s^{-1}}$) than retrograde counterparts (up to $60\,\mathrm{m\,s^{-1}}$) and the overall scheme is qualitatively consistent with the zones/belts circulation described above, with air parcels accelerated along rotation direction while moving toward the poles and accelerated against rotation while moving toward the equator. The entire wind zonal pattern displays marked hemispherical asymmetry, and a variability in the order of 15% in absolute values of wind speed between the Voyager and Cassini era.

Above this general zonal pattern, a large variety of phenomena is also observed. The famous Great Red Spot (GRS) is a large (diameter in the order of 1.6×10^4 km) anticyclonic formation residing at about 20° S, whose existence has been documented for at least two centuries. Its central parts appear dark in the infrared, while it has been demonstrated that altitude is higher (in the order of 10 km) than typically observed on Jupiter. The higher altitude has been invoked as a cause of the red colour, since higher levels of cloud decks would provide sites for chemical catalysis for photodissociation products. A number of other anticyclones with relatively high-altitude cores have been also observed for time scales in the order of several decades, in the form of bright White Ovals (WO). Both GRS and WOs are well seen in Fig. 4.14. Anticyclonic systems are accompanied by extensive turbulent wakes developed in variety of spatial scales. Turbulent phenomena are also evident in several latitude regions characterized by strong latitudinal wind shear.

Other remarkable features are the so-called Hot Spots, elongated regions (extended typically 2×10^3 km in the meridional direction and 10^4 km along the parallels) of very bright IR emissions located at the edge of between the Equatorial Zone and the north Equatorial Belt (Fig. 4.16). Here the main cloud decks become extremely thin, and in the IR range—where haze is essentially transparent—aerosol opacities well below than 1 can be achieved, allowing the study of minor gases at depths of a few bars. Hot spots has also been directly explored by the Galileo Entry Probe (GEP),

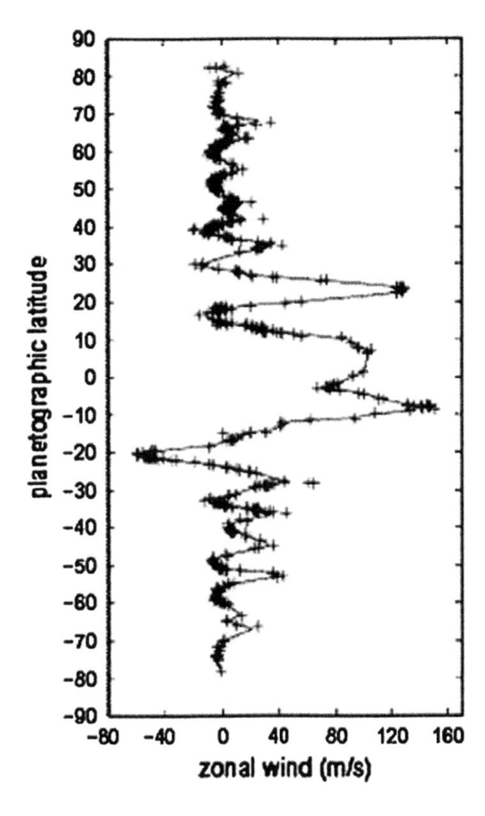

Fig. 4.15 Zonal winds in the Jupiter atmosphere as a function of latitude, as derived from Cassini Imaging Science Subsystem data. Reprinted from Li et al. (2004), Copyright 2004, with permission from Elsevier

proving the only instance at the date of in-situ measurements of a giant planet atmosphere. A notable numerical model (Showman and Dowling 2000) sees the Hot Spots as the results of Rossby Waves, and the complex patterns observed in wind and in minor gases distributions are consistent with this hypothesis.

Further insights on the overall structure of Jupiter circulation is provided by the air temperature longitudinal section (Fletcher et al. 2009), see Fig. 4.17. The zone/belt pattern has clear counterparts in the temperature minima and maxima particularly evident at the 0.1 bar, where downwelling over the zones induces strong air warming. Air temperatures above the 0.1 bar levels tends to increase with altitude, as expected from absorption of solar UV radiation, but the latitudinal trends show a less obvious behaviour and, at low latitudes, seems to be subject to long terms oscillations in the order of four years (quasi-quadrennial oscillations QQO), not consistent with variations in solar forcing. The deeper regions sensed by these retrievals lies very close to the 1 bar level, where longitudinal gradients are minimal. The direct sampling by GEP confirmed at the entry Hot Spot below the 1 bar level a temperature gradient essentially equivalent to the adiabatic at least down to 20 bar.

Latitudinal distribution of minor components of the atmosphere are also significant to complete the available picture. Phosphine, a disequilibrium species that is

Fig. 4.16 Simulated view of Jupiter as observed at 5 μm as reconstructed from the data of the Jupiter Infrared and Auroral Mapper (JIRAM) on board of the Juno spacecraft. Arrow marks the position of a bright Hot Spots. Courtesy by Mura, Adriani, Bolton and the JIRAM/Juno team

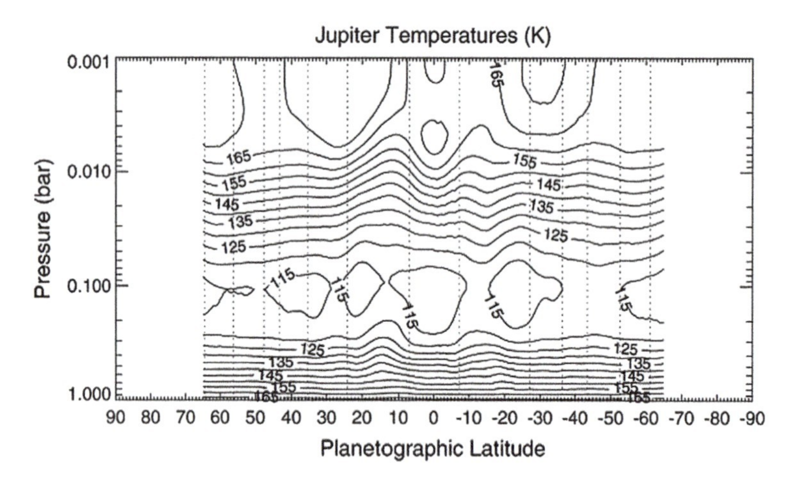

Fig. 4.17 Mean air temperatures in Jupiter upper troposphere and lower stratosphere as a function of altitude and latitude. Values were retrieved from the data of the Cassini Composite Infrared Spectrometer acquired during the Jupiter flyby between December 31, 2000 and January 10, 2001. Reprinted from Fletcher et al. (2009), Copyright 2009, with permission from Elsevier

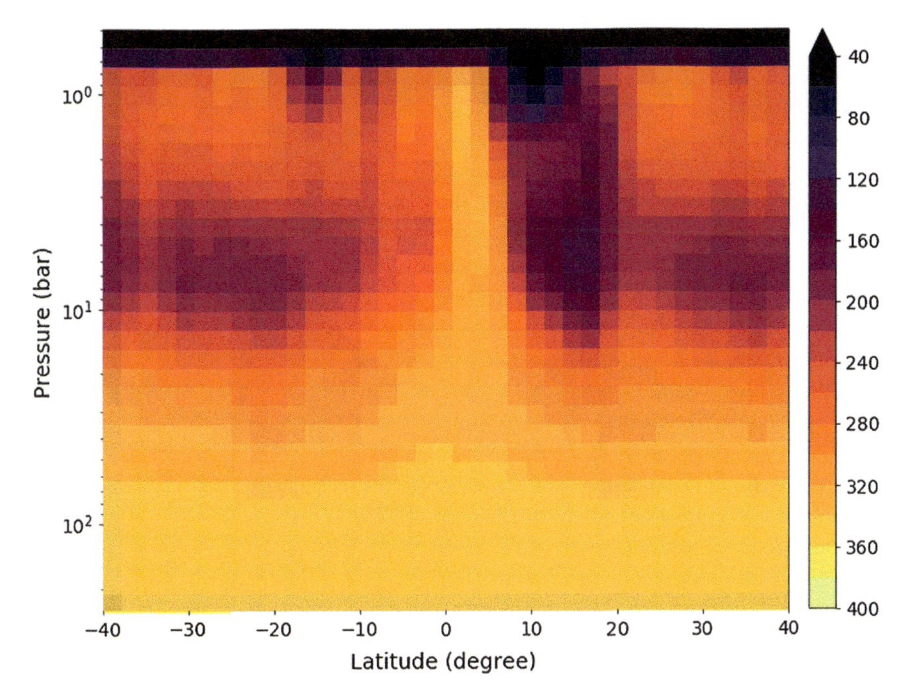

Fig. 4.18 Content of ammonia (in parts per millions over volume) in the Jupiter troposphere, as a function of latitude and pressure, as retrieved from the data of the Juno Microwave Radiometer (MWR) acquired during a single pericenter passage on 27 August 2016. Courtesy by Li et al. (2017). For an updated version of this plot, see Li et al. (2017)

expected to be completely removed by reaction with water vapour for $T < 2000\,K$ and by photodissociation in the upper troposphere, is commonly assumed as a proxy of vertical motions from the deepest parts of the atmosphere. Global maps presented by Fletcher et al. (2009) for levels above the 1 bar shows clear maxima over the Equatorial zone, that is further confirmed as a region of vertical up welling. Recent data from the Juno mission has eventually allowed to track the ammonia content in the 40° S—40° N band down to at least 100 bar (Li et al. 2017), see Fig. 4.18. Retrievals demonstrate the occurrence of a previously unexpected global deep circulation, that creates a plume of ammonia-rich at the location of the Equatorial zone from a deep reservoir below the 50 bar levels. The north Equatorial belt appears as an area of ammonia depletion, but, beside this specific feature, the zone/belts pattern has no clear trace in ammonia maps below the approximate 3 bar level.

4.4.2.2 Saturn

The basic properties of the atmosphere of Saturn are qualitatively similar to the ones previously described for Jupiter. A complete review is provided by Dougherty et al. (2009). A key difference is represented by the enhanced enrichment in heavy

Fig. 4.19 Mosaic of true color images of Saturn from Cassini Imaging Science Subsystem. NASA Photojournal PIA11141. Courtesy NASA/JPL/Space Science Institute

elements (with respect to Solar composition) once compared against the Jupiter case. This is consistent with the overall smaller mass. The original proto-Saturn core of heavier components had a smaller mass than its Jupiter counterpart and was therefore less effective in collecting the lighter gases such as hydrogen and helium. The final result is a lesser degree of dilution of heavier species.

Despite the smaller mass and hence a smaller internal pressure at fixed radius from the center, it is believed that hydrogen is still capable to achieve a metallic state in the deep interior of Saturn, and therefore able to efficiently sequester helium in the deep interior. Albeit the estimates of this noble gas remain rather indirect in the persisting absence of in-situ measurements, latest works suggest a helium abundance closer to Jupiter values than previously assumed.

The energy balance of Saturn is slightly greater than in the Jupiter case, but once the greater distance of the Sun is taken into account, the net energy emitted per unit mass is nevertheless smaller, consistently with the higher surface/volume ratio.

The cloud structure is very similar to the one expected for Jupiter, but shifted downward at higher pressure levels because of lower air temperatures. Sub-micron hazes in the upper troposphere/lower stratosphere are particularly thick over the equatorial region, where they reach their higher altitude. Here, visible optical depth exceeds 0.8, to fall below 0.05 beyond 60° latitude. The combined effects of deeper clouds (with enhancement of Rayleigh scattering) and thicker hazes reduces considerably the contrast of atmospheric features as observed in the visible spectral range once compared against Jupiter (Fig. 4.19). Similar to the Jupiter case, peculiar stratospheric haze has been observed in polar regions.

A clear zonal structure of the atmosphere exists also for Saturn. Corresponding zonal jet speed reaches values up to $400\,\mathrm{m\,s^{-1}}$ at about 10° latitude from the equator (Fig. 4.20). These high winds are effective in dissipating the larger cyclonic systems, observed on Saturn but lacking the long lifetimes of their Jupiter counterparts.

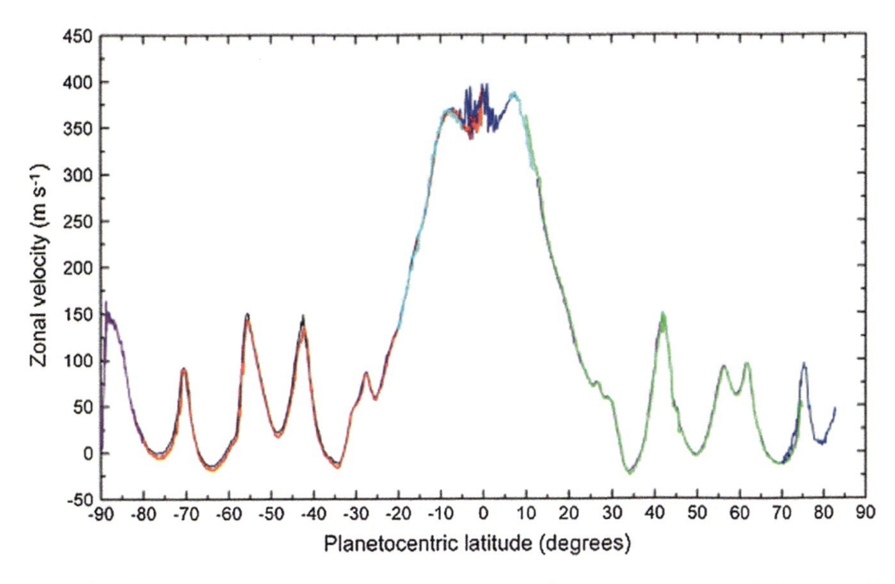

Fig. 4.20 Zonal winds in the upper troposphere of Saturn. Reprinted from García-Melendo et al. (2011), Copyright 2011, with permission from Elsevier

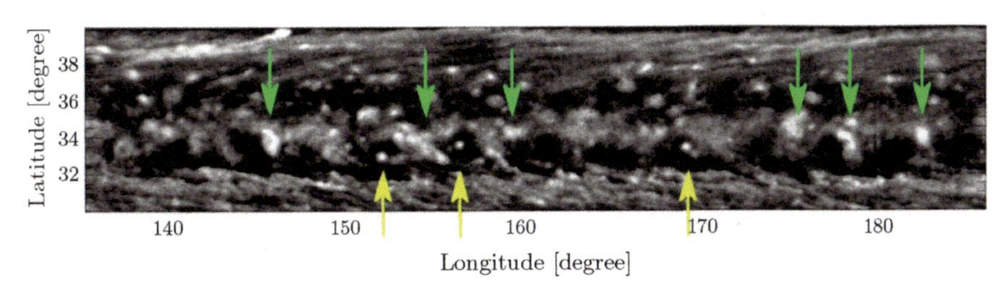

Fig. 4.21 A chain of IR-bright cyclones in the troposphere of Saturn – the "String of Pearls"—appears as a series of dark spots at the approximate latitude of 33° N in this Cassini Imaging Science Subsystem image. Reprinted from Sayanagi et al. (2014), Copyright 2014, with permission from Elsevier

IR observations by the Cassini Spacecraft have revealed both the very fine structure of the zonal pattern as well as a large variety of small vortex features (such as e.g.: annular clouds or the "String of Pearl", Fig. 4.21).

Among other phenomena of the Saturn atmosphere, we remind here the north pole hexagon (Fig. 4.22), a cloud feature that exhibits a rigid rotation of its sides (with a length in the order of 1.3×10^4 km) at an angular speed equal to the one inferred for global magnetic field. No consensus has yet been achieved about its origin: the original suggestion of a Rossby wave have been further elaborated to include a number of non-linear effects. Differential fluid rotation has also been invoked as a possible ultimate cause. The absence of a similar structure in the south pole and its persistence since the Voyager epoch to the entire Cassini lifetime suggests indeed a

Fig. 4.22 The north pole of Saturn as seen by the Cassini Imaging Science Subsystem image. NASA Photojournal PIA18274. Courtesy NASA/JPL-Caltech/Space Science Institute

critical dependence upon local properties of the atmosphere. The area immediately over the pole and well inside the hexagon shows an anticyclonic circulation, similarly to what observed on Jupiter. The south pole of Saturn hosts a large vortex structure, with a notable wall of clouds towering over a central "eye" where aerosol reside at much lower altitude.

Large-scale transient phenomena have been recorded since long time by telescopic observers. A notable case was documented with unprecedented detail by the Cassini spacecraft between 2010 and 2011 (Sayanagi et al. 2013). In that instance, a large convective outburst developed in a time scale of a month, encircling the entire latitudinal band around 33° N within a month. The activity led to the formation, among other features, of an anticyclone with a size in the order of 1.1×10^4 km. The activity remained detectable for at least six months, with long term variations on cloud albedo and zonal wind profiles of affected areas.

Air temperature fields of Saturn upper troposphere bear clear trace of the zonal structure (Fletcher et al. 2009; Fig. 4.23), with maximum variations at 7×10^{-2} bar.

Longitudinal gradient is extremely weak below 0.5 bar. Because of axial tilt, air temperatures above the troposphere shows a marked hemispherical asymmetry, consistent with different degrees of solar forcing. The upper troposphere (0.2 bar) is characterized by warming over both polar regions.

In absence of extensive microwave observations, constraints on Saturn deep circulation remains rather indirect. Phosphine presents a major rise on the equator as in the Jupiter case, suggesting therefore a strong upwelling from deeper layers at these locations. On the Saturn case however, rises of comparable magnitude are also detected at least in other three southern latitude bands.

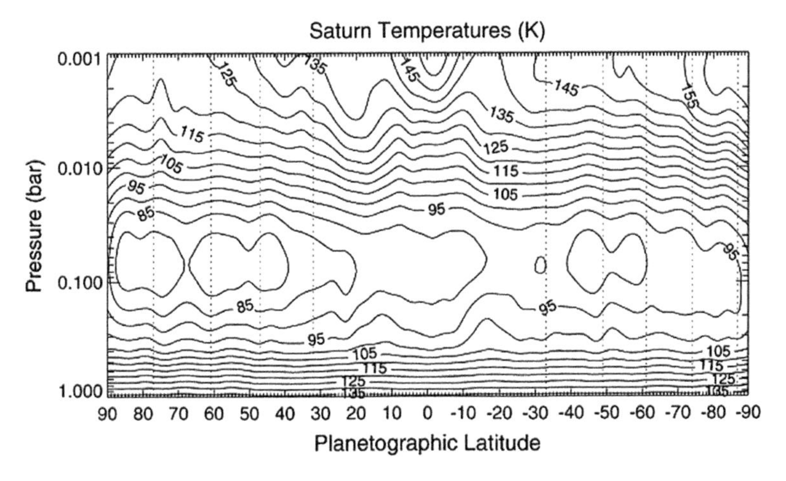

Fig. 4.23 Mean air temperatures in Saturn upper troposphere and lower stratosphere as a function of altitude and latitude. Values were retrieved from the data of the Cassini Composite Infrared Spectrometer. Reprinted from Fletcher et al. (2009), Copyright 2009, with permission from Elsevier

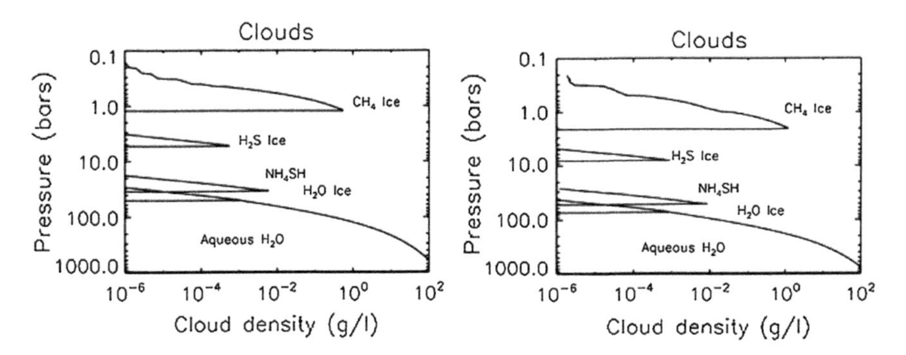

Fig. 4.24 Theoretical profiles of aerosol concentrations in the atmospheres of Uranus (left-hand panel) and Neptune (right-hand panel). From Irwin (2009)

4.4.2.3 Uranus and Neptune

Uranus and Neptune show some fundamental differences against Jupiter and Saturn. The smaller masses of initial cores did not allow the build-up of extensive gaseous envelopes in the later phases of formation. Consequently, the overall compositions - albeit still dominated by hydrogen and helium in the outer mantles - contains seizable amounts of heavier components, brought to the bodies in the form of volatiles-rich icy planetesimal, justifying therefore the class name of Icy Giants. The smaller internal pressure and different composition did not allow the creation of extensive mantles of metallic hydrogen in the deep interior. This is consistent with magnetic-field data, that contain important high orders components, contrarily to Jupiter and Saturn, largely dominated by the dipolar term.

Despite their apparent similarity, Uranus and Neptune present some evident differences, that point toward rather different structure of planetary deep interiors. Uranus density is about 20% lower than Neptune, leading eventually to a larger radius. Moreover, while Uranus appears to be essentially in energy balance, Neptune shows a clear energy excess (albeit with a power emitted per mass unit about five time smaller than gas giants).

The outermost cloud systems of icy giants present some basic difference against Jupiter and Saturn (Fig. 4.24). On the basis of microwave observations, hydrogen sulphide is expected to be in excess against ammonia on icy giants. This latter species is completely removed in formation of the ammonium hydro sulphide cloud deck, while the residual hydrogen sulphide forms ice clouds at slightly higher altitudes. More important, the upper tropospheres of both planets are so cold that methane is allowed to freeze out. In these conditions, the hydrogen-helium atmosphere remains largely depleted of all minor components, excluding noble gases. In matter of fact,

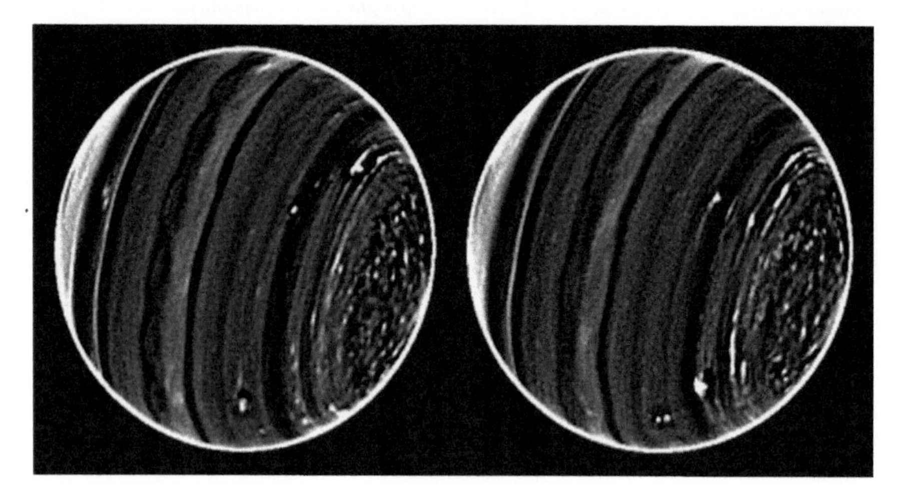

Fig. 4.25 Contrast-enhanced images of Uranus as observed by the Keck telescope in the H-band (1.65 μm). Reprinted from Sromovsky et al. (2015), Copyright 2015, with permission from Elsevier

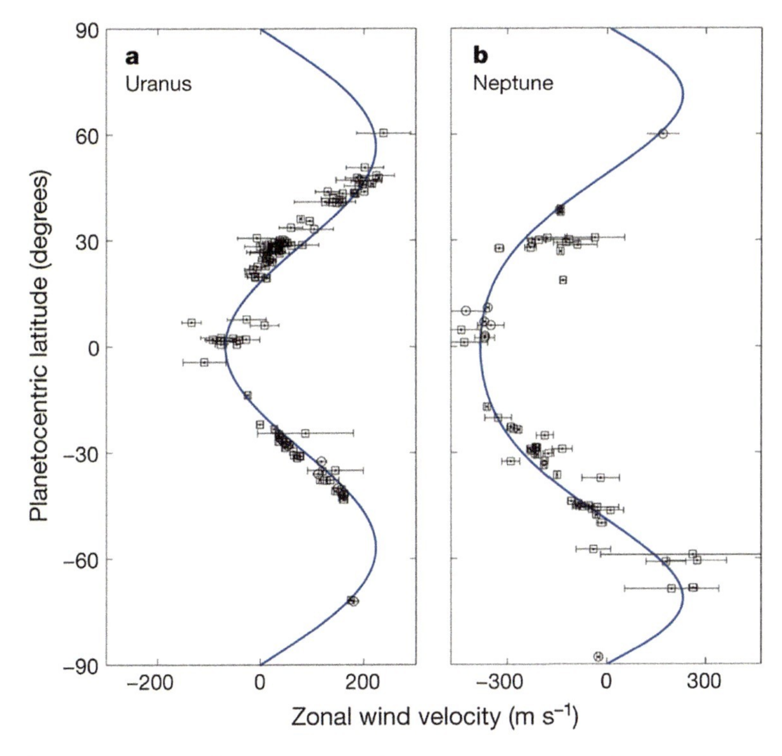

Fig. 4.26 Zonal wind speeds on Uranus and Neptune, as retrieved from Voyager 2 (circles) and HST (squares) data. Reprinted by permission from Macmillan Publishers Ltd: Nature, Kaspi et al. (2013), Copyright 2013

positive detection of hydrocarbons in the stratosphere (produced by photodissociation of methane) suggest the occurrence of vertical motions capable to replenish the stratosphere. Given the higher amounts detected on Neptune, this planet is expected to host more vigorous vertical motions in comparison against Uranus.

Both planets present a clear zonal patterns of clouds, that in the Uranus case become particularly evident in images from adaptive-optic ground-based telescopes. On the Uranus there is a clear disruption of zonal pattern beyond about 60° S, given the patchy appearance of the entire polar region (Fig. 4.25). The retrieved zonal wind fields present two, hemisphere symmetric, prograde jets at about 60° latitude and a retrograde jet on the equator (Fig. 4.26). Jets are strong: on Neptune they exceed 300 m s^{-1}, the higher values reported to the date in the entire solar System. Available information suggests that wind patterns and thermal structure of the upper troposphere of Uranus were subject only to minor changes from the Voyager era to at least 2011 (Orton et al. 2015), despite the strong seasonal variations caused by the extreme axial tilt of the planet. The structure of wind fields is consistent with the available Uranus air temperatures latitudinal cross section: at 0.1 bar, two hemispheric symmetric minima are found at 40° N and 40° S, with maxima of comparable amplitude

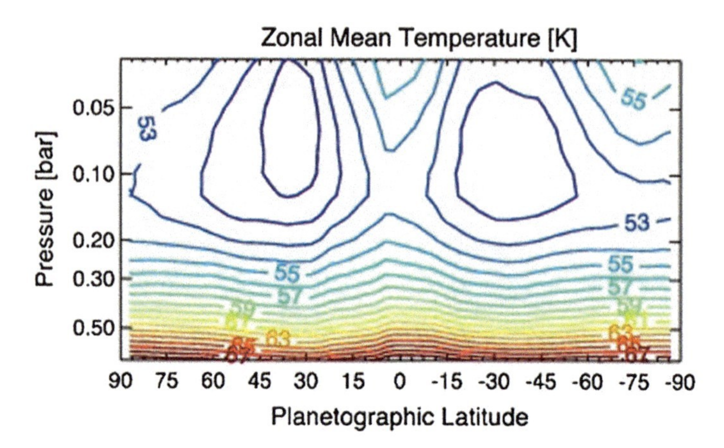

Fig. 4.27 Mean air temperature in the Uranus upper troposphere/lower stratosphere. Reprinted from Orton et al. (2015), Copyright 2015, with permission from Elsevier

at the equator and over both poles (Fig. 4.27). These data are overall consistent with upwelling at intermediate latitudes and subsidence at high latitudes and at the equator. However, this is contrast with the global deep (few tents of bar) distributions of ammonia inferred from microwave ground observations, that suggest an upwelling at equator from deep interior (as seen on Jupiter) and subsidence at the poles.

Icy-giant atmospheres are characterized by a rich phenomenology. Albeit Uranus shows very little details during the 1986 Voyager encounter (occurred in the vicinity of the summer solstice), atmospheric phenomena (mostly in the form of distinctive bright high-altitude clouds, presumably formed of methane ice) has been observed more and more frequently moving toward the northern spring equinox in 2008 and in subsequent years (de Pater et al. 2015). A bright polar cap on the southern hemisphere has progressively disappeared while moving toward the equinox. Another notable, transient feature of Uranus was the Uranus Dark Spot, a darker area observed at intermediate latitudes in 2006 with a size of about 1500 km (Fig. 4.28). A hypothesis on its nature sees it as the result of an anticyclonic system, creating an area of depleted cloud coverage at its center.

Neptune displayed a richer appearance already during the Voyager encounter (Fig. 4.29). The abrupt appearance of very bright clouds has been documented continuously after initial observations from Hubble Space Telescope. These cloud systems appear often much larger, brighter and longitudinally extended than their counterparts in Uranus (Fig. 4.30). A Great Dark Spot was observed in details by Voyager, about five times larger than Uranus Dark Spot described above. It was surrounded by brighter high-altitude clouds and completely disappeared before Hubble Neptune observations in 1994. Another Neptune dark spot has been observed since 2015 in the southern hemisphere, again accompanied by brighter high clouds. As in the Uranus case, these features are interpreted as regions of cloud clearance that expose deeper atmospheric layers. These locations, in view of the possibility that they offer to penetrate deeper in the troposphere of Icy Giants, will represent important sites for the observations of future missions to these remote planets (Turrini et al. 2014).

4.4.3 Icy Bodies

The outer Solar System hosts a large number of solid bodies characterized by external outer surfaces composed mostly of ices, being water ice by far the most important component. Table 4.5 summarizes the key properties of the atmospheres of these icy bodies.

In most circumstances, these bodies are surrounded only by tenuous exospheres, generated by the interaction of the surfaces with space environment. For example, sputtering by charged particles accelerated in the Jovian magnetosphere is the primary mechanism generating the exosphere of Europa (Plainaki et al. 2012). There are however some cases where conditions are such to create nitrogen-rich atmospheres fully in the collisional regimes.

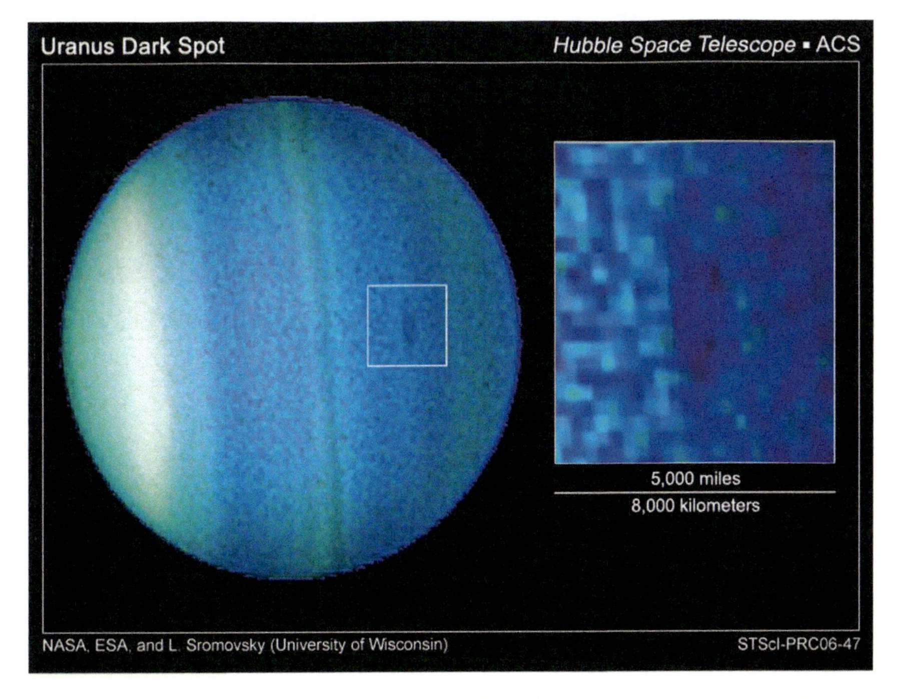

Fig. 4.28 The Uranus Dark Spot, as seen by HST. STScI release PRC-06-47. Courtesy NASA/ESA/Sromovsky (Univ. of Wisconsin)

Fig. 4.29 Neptune as seen by the Voyager 2 Narrow Angle Camera. At the centre of the image, the Great Dark Spot. NASA Photojournal PIA01492. Courtesy NASA/JPL

4.4.3.1 Titan

The Saturn moon Titan represents a unique case in the Solar System. It exhibits the thickest atmosphere surrounding a moon or an icy body (Fig. 4.31). Moreover, the complex interactions between surface, interior and spatial environment are excellent examples of how apparently minor processes can lead to substantial differences in planetary evolution.

The equilibrium temperature at Titan (85 K) is very close to the freezing point of molecular nitrogen (65 K), the main component of the atmosphere. The greenhouse effect caused by methane (and by its dissociation products) is therefore essential to maintain the entire atmosphere in place, by rising the mean temperature to about 93 K. On the other hand, methane is prone to photochemical dissociation and, at the current date, no evident mechanism to cycle it back from its dissociation products

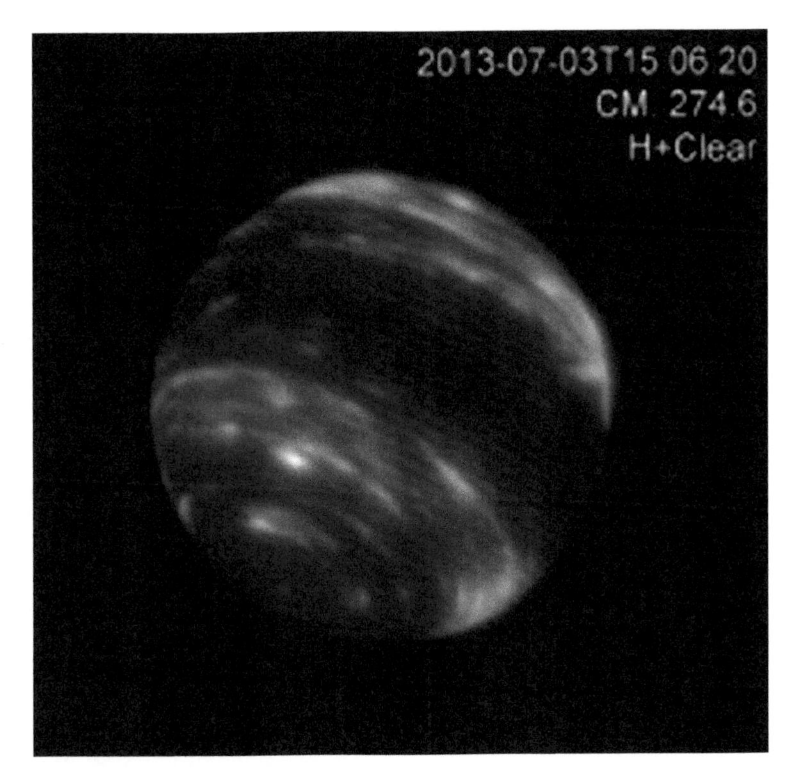

Fig. 4.30 Neptune as observed by the Keck telescope in the H-band (1.65 μm). Reprinted from Hueso et al. (2017), Copyright 2017, with permission from Elsevier

Table 4.5 Properties of icy bodies more relevant for the discussion on their atmospheres. Updated from NASA Planetary Fact Sheets (https://nssdc.gsfc.nasa.gov/planetary/factsheet/)

Parameter	Unit	Titan	Triton and Pluto (Kuiper-belt objects)	Jupiter icy moons
Mass of the body	$M_\oplus$	0.0225	0.0022 (Pluto) 0.0036 (Triton)	0.008 (Europa) 0.025 (Ganymede) 0.018 (Callisto)
Orbital semi-major axis	au	9.5	39.4($e = 0.2$) (Pluto) 30.1 (Triton)	5.2
Sideral period	days	15.95	6.38 (Pluto) 5.8 (Triton)	3.5 (Europa) 7.1 (Ganymede) 16.7 (Callisto)
Surf. pressure	bar	1.5	1.0×10^{-5} (Pluto) 1.4×10^{-5} (Triton)	1×10^{-12} (Europa) $1 - 10 \times 10^{-12}$ (Ganym.)
Axial Tilt	degree	23.45	122.5 (Pluto)	1.5
Main components	Volume fraction	N_2 (0.98), CH_4	N_2 (0.99), CH_4 (0.01), N_2 (0.99), CO	H_2O, O_2, S, H_2

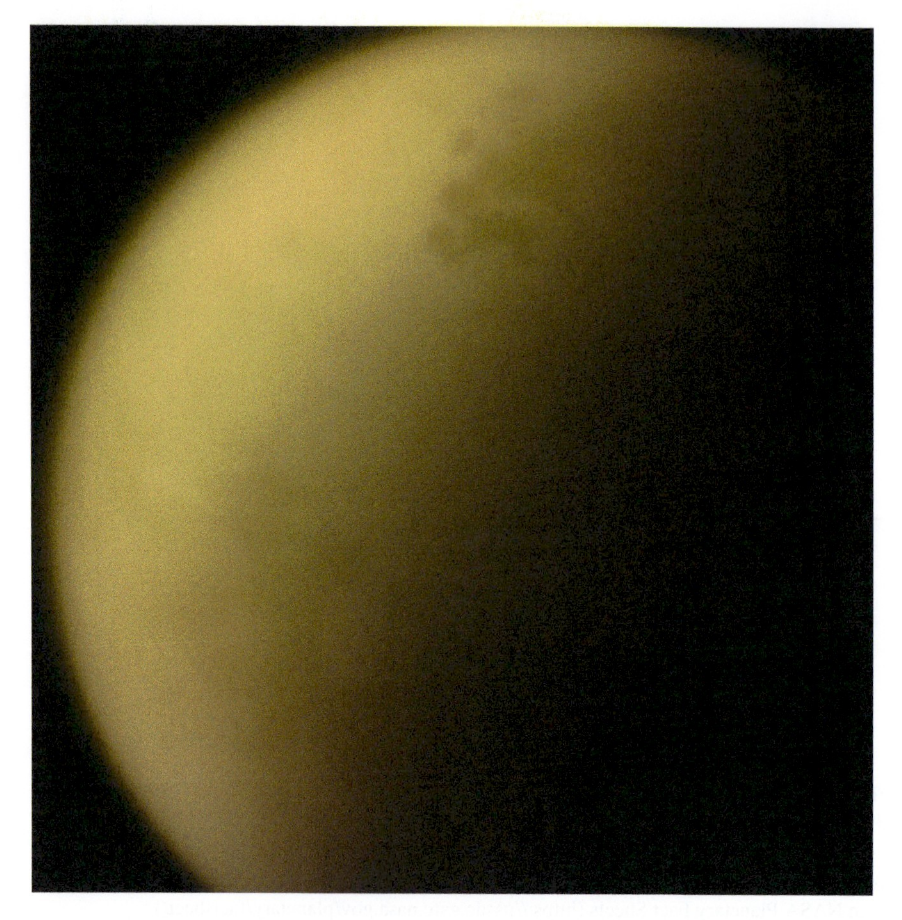

Fig. 4.31 Natural colour view of Titan as observed by the Cassini Narrow Angle Camera. NASA Photojournal PIA21890. Courtesy NASA/JPL-Caltech/Space Science Institute

has been identify. It is therefore necessary to invoke some mechanism of continuous replenishment over geological time scales.

A possible scenario points to the deep water ocean beneath the Titan crust (mostly composed by water ice and ammonia) as the ultimate methane source. During the early stages of Titan evolution, a deep ocean must have been directly in contact with the rocky inner core. In these conditions, and in presence of mafic rocks (such as olivine and pyroxenes) and carbon-bearing species (such as carbon dioxide dissolved in the water), serpentine minerals and methane are produced (Atreya et al. 2006). Models predict that at the current date a layer of ice separates liquid water from rocks, inhibiting therefore further synthesis of methane. The gas previously produced is however effectively trapped in the deep ice layers as *clathrates* (peculiar crystalline forms of ice capable to encapsulate gas molecules in the gaps between water molecules) and is released slowly over geological times. Ultimately, methane diffuses from the deep ocean into the atmosphere through cracks in the upper crust. The diffusion scenario is supported by the positive detection of Ar^{40}, an isotope produced by decay of K^{40} occurring in the rocky core.

Once released, the methane enables one of the most complex chemical cycle documented in the Solar System. Its dissociation by UV photolysis occurs along dissociation of N_2 by impinging of charged particles accelerated by the Saturnian magnetosphere. The products are prone to complex cycles to produce eventually complex molecules such as nitriles and polycyclic aromatic hydrocarbons. These molecules are the main components of high altitude diffuse sub-micron hazes (about 70 km over the surface) that complicate the visual observations of the surface. Other notable products of methane photodissociation are ethane and propane, which may exist in stable liquid form in certain locations of Titan surface. Cassini instruments were indeed capable to detect at both Titan poles bodies of liquid in the form of large and relatively shallow lakes (Fig. 4.32). River and coastal landforms, albeit currently dry, have also been documented at intermediate latitudes. Titan hosts therefore the only currently active hydrological cycle (based on hydrocarbons) known in the Solar System (beside the Earth water cycle).

Low altitude clouds of methane and molecular nitrogen condensing at the tropopause allowed to constrain the dynamic of the atmosphere (Fig. 4.33). The Titan atmosphere shows a super rotation overtaking the solid-body rotation period of 16 Earth days, similar to what observed in Venus. Around equinoxes, the circulation pattern of the atmosphere is organized according two roughly symmetric Hadley cells (Lebonnois et al. 2012). This is consistent with extensive (methane) cloud system observed along the equator in this period. However, given the high effective axial tilt, Titan is subject to strong seasonal variations along the Saturnian year. These include variations of lakes amplitudes and the development of thick polar hoods over the winter pole.

4.4.3.2 Pluto and Triton

Kupier-Belt Objects are the last class of Solar System objects with substantial fraction of their atmospheres in the collisional regime.

Composition of their atmospheres is dominated by molecular nitrogen and methane, with both species in equilibrium with their ices at the surface. Both bodies have marked seasonal cycles, caused by axial tilts and, in the case of Pluto, by strong orbital eccentricity. On Pluto, long-term ground measurements have actually demonstrated a variability of total surface pressure in the order of 60% along the orbit. This is considered as a proxy of seasonal migration of more volatile ices and indeed nitrogen, methane and carbon monoxide areas spectroscopically detected by New Horizon consistently appear brighter and of fresher appearance than stable water ice rich regions (Fig. 4.34). Morphological and albedo differences observed between

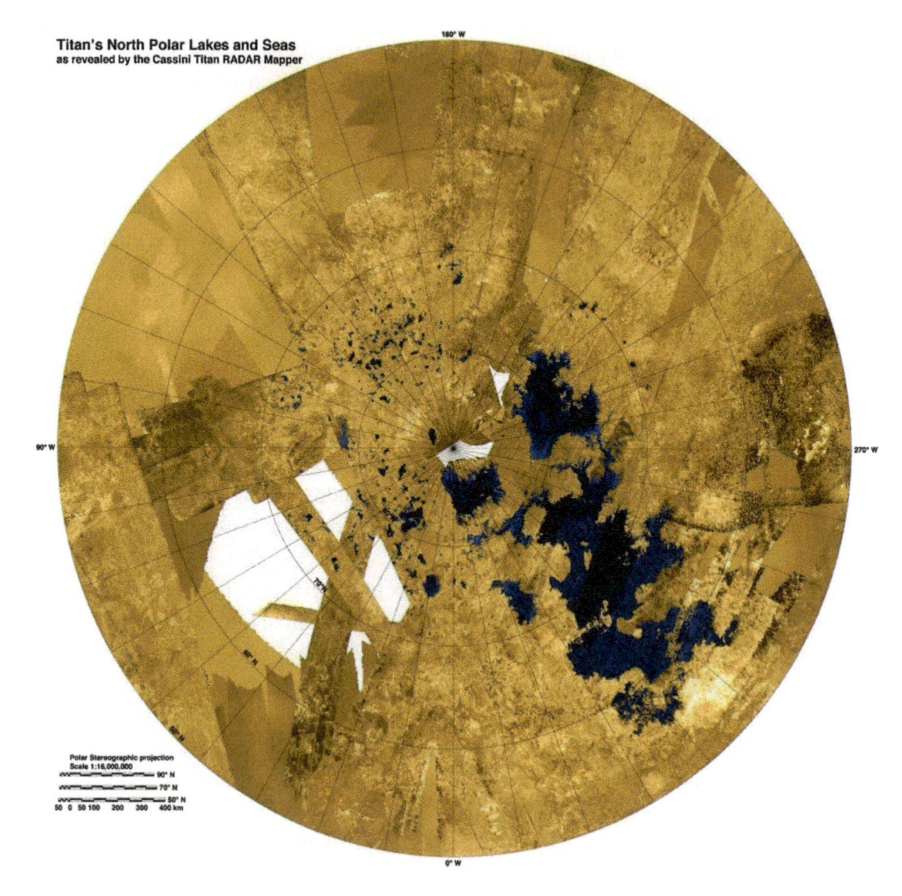

Fig. 4.32 Map of hydrocarbon lakes in the Titan north polar region, as revealed by Cassini radar. NASA Photojournal PIA 17655. Courtesy NASA/JPL-Caltech/ASI/USGS

equatorial and polar regions of Triton has similarly been interpreted as due to a song term cycle of ices sublimation/deposition.

Measured Pluto temperature profile does not show any detectable troposphere (Fig. 4.35). The lowest atmosphere is indeed characterized by an increase of temperature with altitude, caused by the methane-induced greenhouse effect. A peak temperature is reached at about 30 km above the surface (110 K), followed by a much smoother decrease associated to the effective cooling by IR emission by CO. Limb images clearly demonstrated the existence of several distinct layers of aerosols in the lowest 100 km of the atmosphere, likely formed by nitrogen and methane (Fig. 4.36). Cloud has been unambiguously identified also on the limb of Triton (Fig. 4.37).

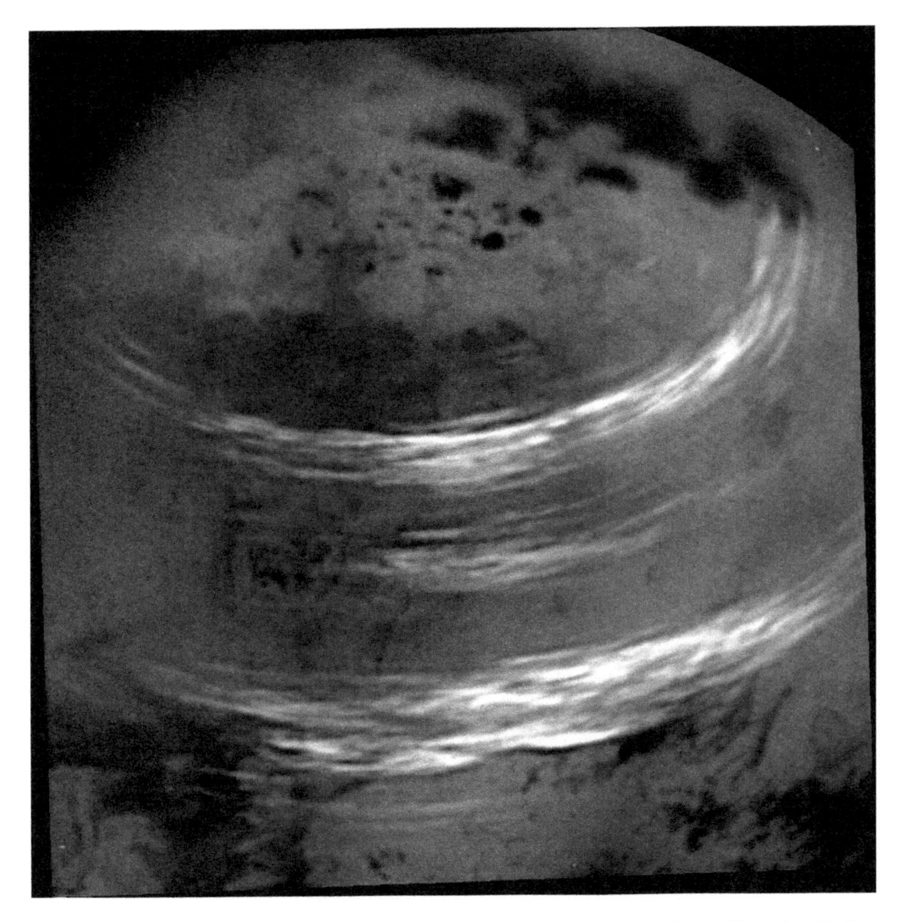

Fig. 4.33 Contrast-enhanced image of methane clouds over the northern Titan hemisphere by the Cassini Imaging Science Subsystem. NASA Photojournal PIA21450. Courtesy NASA/JPLCaltech/Space Science Institute

Acetylene and ethylene mixing ratios vs altitude have been measured on Pluto. An effective methane dissociation by UV at these large heliocentric distances is not expected, and therefore a direct action from solar wind has been invoked to justify observed abundances of these minor species. Despite the very low pressure, dark material observed on the darker (and most stable) Pluto regions is believed to be formed by complex organic molecules ultimately derived from dissociation of methane and nitrogen.

Fig. 4.34 Distribution of different types of ices on the surface of Pluto. *Top panel:* CH_4, N_2, and CO. *Bottom panel:* H_2O. Reprinted with permission from AAAS. From (Grundy et al. 2016)

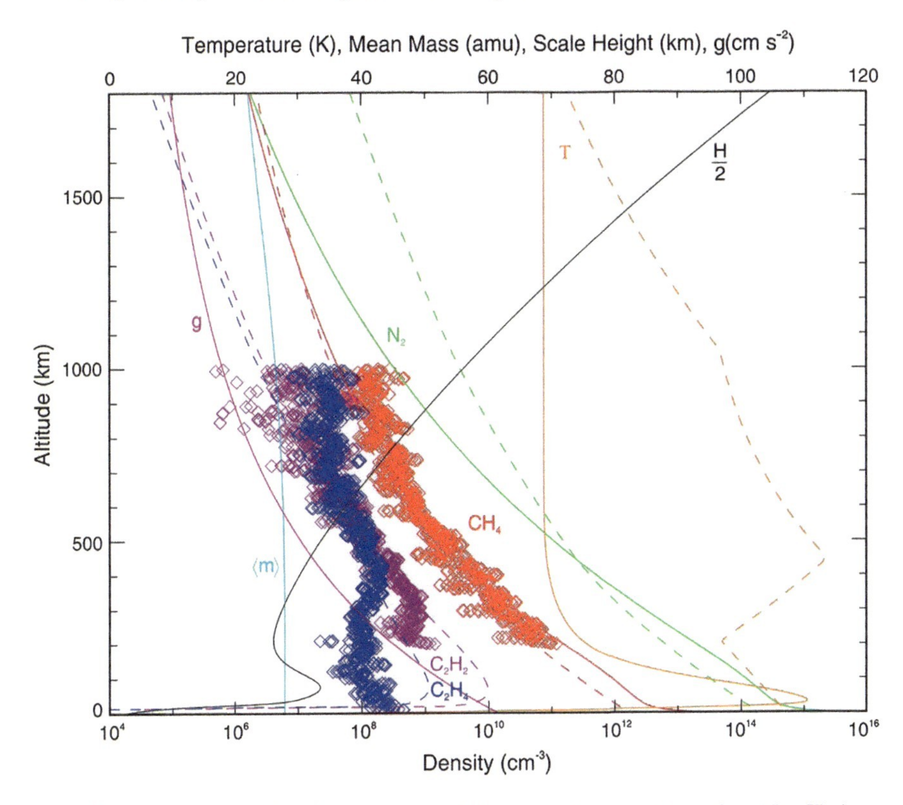

Fig. 4.35 Temperature profile and composition of Pluto atmosphere as inferred from New Horizons data. Reprinted with permission from AAAS. From Gladstone et al. (2016)

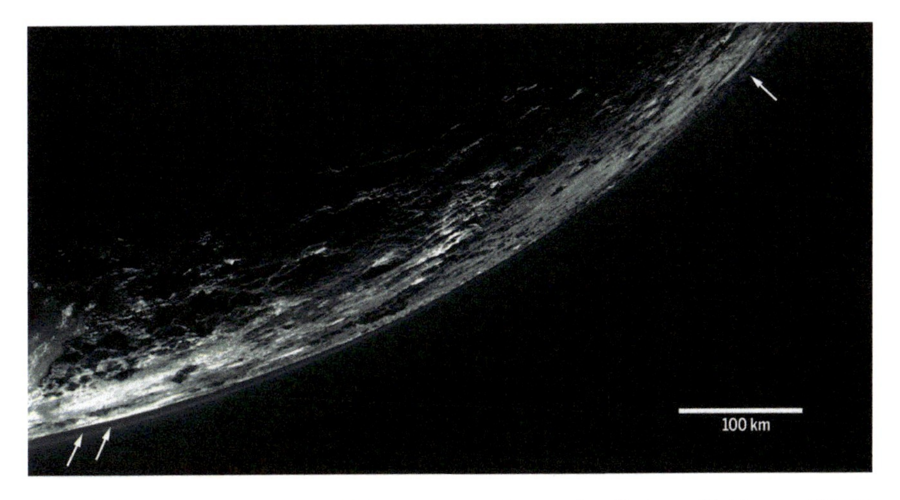

Fig. 4.36 Hazes on over the limb of Pluto, as observed by the New Horizons Multispectral Visible Imaging Camera (MVIC). Reprinted with permission from AAAS. From Gladstone et al. (2016)

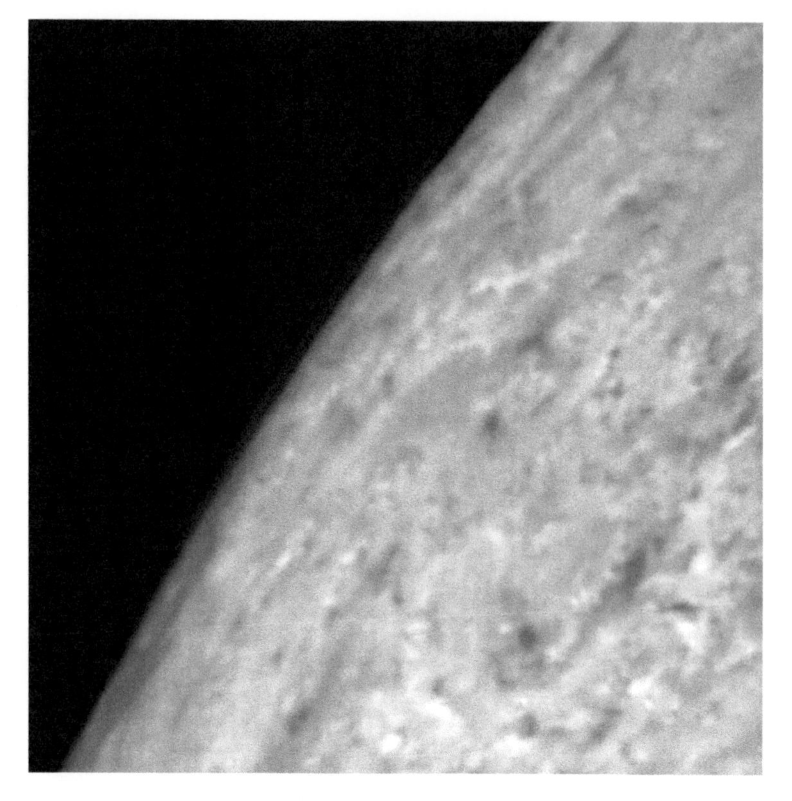

Fig. 4.37 Hazes on over the limb of Triton, as observed by the Voyager 2 Imaging Science System (ISS). NASA Photojournal PIA02203

References

Altwegg, K., Balsiger, H., Bar-Nun, A., et al.: Science **347**, 6220 (2015)

Atreya, S.K., Wong, M.H., Owen, T.C., Mahaffy, P.R., Niemann, H.B., de Pater, I., Drossart, P., Encrenaz, Th: Planet. Space Sci. **47**, 1243 (1999)

Atreya, S.K., Adams, E.Y., Niemann, H.B., et al.: Planet. Space Sci. **54**, 1177 (2006)

Atreya, S.K., Crida, A., Guillot, T., Lunine, J.I., Madhusudhan, N., Mousis, O.: in Baines, K., Flasar, M., Krupp, N., Stallard, T. (eds.) Saturn in the 21st Century. Cambridge University Press, in press (2018). arXiv:1606.04510v2

Bagenal, F., Dowling, T.E., McKinnon, W.B.: Jupiter. The Planet, Satellites and Magnetosphere. Cambridge University Press (2004)

Bird, G.A.: Phys. Fluids **13**, 2676 (1970)

Boynton, W.V., Feldman, W.C., Squyres, S.W., et al.: Science **297**, 5578 (2002)

Carlson, R.W., Baines, K.H., Anderson, M.S., Filacchione, G., Simon, A.A.: Icarus **274**, 106 (2016)

Curry, S., Luhmann, J., Dong, C., et al.: Am. Astron. Soc. DPS meeting #49, id.#510.02 (2017)

de Pater, I., Lissauer, J.J.: Planetary Sciences. Cambridge University Press (2010)

de Pater, I., Sromovsky, L.A., Fry, P.M., Hammel, H.B., Baranec, C., Sayanagi, K.M.: Icarus **252**, 121 (2015)

Dougherty, M., Esposito, L., Krimigis, S.: Saturn from Cassini-Huygens. Springer (2009)

Falkowski, P., Scholes, R.J., Boyle, E., et al.: Science **290**, 5490 (2000)
Fedorov, A., Barabash, S., Sauvaud, J.A., Futaana, Y., Zhang, T.L., Lundin, R., Ferrier, C.: J. Geophys. Res. **116**, A07220 (2011)
Fletcher, L.N., Orton, G.S., Teanby, N.A., Irwin, P.G.J.: Icarus **202**, 543 (2009)
Fukuhara, T., Futaguchi, M., Hashimoto, G.L., et al.: Nat. Geosci. **10**, 85 (2017)
García-Melendo, E., Pérez-Hoyos, S., Sánchez-Lavega, A., Hueso, R.: Icarus **215**, 62 (2011)
Gladstone, G.R., Stern, S.A., Ennico, K., et al.: Science **351**, aad8866 (2016)
Grundy, W.M., Binzel, R.P., Buratti, B.J., et al.: Science **351**, aad9189 (2016)
Hanel, R.A, Conrath, B.J., Jennings, D.E., Samuelson, R.E.: Exploration of the Solar System by Infrared Remote Sensing. Cambridge University Press (2003)
Haus, R., Kappel, D., Arnold, G.: Icarus **232**, 232 (2014)
Hueso, R., de Pater, I., Simon, A., et al.: Icarus **295**, 89 (2017)
Ingersoll, A.P.: J. Atmos. Sci. **26**, 6 (1969)
Irwin, P.: Giant Planets of Our Solar System: Atmospheres, Composition, and Structure. Springer (2009)
Jakosky, B.M., Slipski, M., Benna, M., Mahaffy, P., Elrod, M., Yelle, R., Stone, S., Alsaeed, N.: Science **355**, 6332 (2017)
Kaspi, Y., Showman, A.P., Hubbard, W.B., Aharonson, O., Helled, R.: Nature **497**, 344 (2013)
Lebonnois, S., Burgalat, J., Rannou, P., Charnay, B.: Icarus **218**, 707 (2012)
Li, L., Ingersoll, A.P., Vasavada, A.R., Porco, C.C., Del Genio, A.D., Ewald, S.P.: Icarus **172**, 9 (2004)
Li, C., Ingersoll, A., Janssen, M., et al.: Geophys. Res. Lett. **44**, 5317 (2017)
Limaye, S.S., Lebonnois, S., Mahieux, A., et al.: Icarus **294**, 124 (2017)
Määttänen, A., Montmessin, F., Gondet, B., et al.: Icarus **209**, 452 (2010)
McCleese, D.J., Heavens, N.G., Schofield, J.T., et al.: J. Geophys. Res. **115**, E12016 (2010)
Montmessin, F., Lefèvre, F.: Nat. Geosci. **6**, 930 (2013)
Morbidelli, A., Chambers, J., Lunine, J.I., Petit, J.M., Robert, F., Valsecchi, G.B., Cyr, K.E.: Meteor. Planet. Sci. **35**, 6 (2000)
Orton, G.S., Fletcher, L.N., Encrenaz, T., et al.: Icarus **260**, 94 (2015)
Pareschi, L.: Monte Carlo methods for kinetic equations 1: Kinetic equations and their computational challenges. In: KT2009: Tutorials IPAM (2009). http://helper.ipam.ucla.edu/publications/kttut/kttut_8441.pdf
Piccialli, A., Tellmann, S., Titov, D.V., Limaye, S.S., Khatuntsev, I.V., Pätzold, M., Häusler, B.: Icarus **217**, 669 (2012)
Piccioni, G., Drossart, P., Sanchez-Lavega, A., et al.: Nature **450**, 637 (2007)
Plainaki, C., Milillo, A., Mura, A., Orsini, S., Massetti, S., Cassidy, T.: Icarus **218**, 956 (2012)
Robinson, T.D., Catling, D. 5C.: Nat. Geosci. **7**, 12 (2014)
Salby, M.L.: Fundamentals of Atmospheric Physics. Academic Press (1996)
Sayanagi, K.M., Dyudina, U.A., Ewald, S.P., et al.: Icarus **223**, 460 (2013)
Sayanagi, K.M., Dyudina, U.A., Ewald, S.P., et al.: Icarus **229**, 170 (2014)
Shematovich, V.I., Johnson, R.E., Cooper, J.F., Wong, M.C.: Icarus **173**, 2 (2005)
Showman, A.P., Dowling, T.E.: Science **289**, 1737 (2000)
Smith, M.D.: Icarus **167**, 148 (2004)
Smith, M.D., Lapedes, A.S., de Jong, J.C., et al.: Science **306**, 371 (2004)
Sromovsky, L.A., de Pater, I., Fry, P.M., Hammel, H.B., Marcus, P.: Icarus **258**, 192 (2015)
Svedhem, H., Titov, D.V., Taylor, F.W., Witasse, O.: Nature **450**, 629 (2007)
Symonds, R.B., Rose, W.I., Bluth, G.J.S., Gerlach, M.: Rev. Mineral. Geochem. **30**, 1 (1994)
Titov, D.V., Taylor, F.W., Svedhem, H., et al.: Nature **456**, 620 (2008)
Titov, D.V., et al.: in Bezard, B. (ed.) Venus III. Cambridge University Press (2016)
Turrini, D., Politi, R., Peron, R., et al.: **104**, 93 (2014)
Webster, C.R., Mahaffy, P.R., Atreya, S.K., et al.: Science **347**, 415 (2015)
Yung, Y.L., de More, W.B.: Photochemistry of Planetary Atmospheres. Oxford University Press (1999)